Homemade Lightning

Homemade Lightning
Creative Experiments in Electricity

R. A. Ford

Third Edition

McGraw-Hill

New York Chicago San Francisco Lisbon London Madrid
Mexico City Milan New Delhi San Juan Seoul
Singapore Sydney Toronto

Library of Congress Cataloging-in-Publication Data

Ford, R. A.
    Homemade lightning : creative experiments in electricity / R. A. Ford—3d ed.
      p. cm.
    ISBN 0-07-137323-3
    1. Electrostatic apparatus and appliances—Experiments. 2. Electric
generators—Experiments. I. Title.

QC573 .F67 2001
621.31'3—dc21   2001041014

McGraw-Hill

A Division of The McGraw·Hill Companies

2 3 4 5 6 7 8 9 0 DOC/DOC 0 7 6 5 4 3 2

ISBN 0-07-137323-3

*The sponsoring editor for this book was Scott Grillo, the editing
supervisor was Caroline Levine, and the production supervisor was
Sherri Souffrance. It was set in Century Schoolbook by Victoria
Khavkina of McGraw-Hill's Professional Book Group composition unit.*

Printed and bound by R. R. Donnelley & Sons Company.

McGraw-Hill books are available at special quantity discounts to use as
premiums and sales promotions, or for use in corporate training programs. For
more information, please write to the Director of Special Sales, Professional
Publishing, McGraw-Hill, Two Penn Plaza, New York, NY 10121-2298. Or contact
your local bookstore.

Contents

Introduction

Years ago, during my last year in high school, I decided to save up and buy a high-voltage Van de Graaff generator in kit form. I assembled and experimented with this machine, which had a spark potential of 500,000 volts. In the dry Connecticut winters it produced a 17-inch spark every second between its two large terminals. Many weekends were spent experimenting with these strange electrical forces and building capacitors for hotter sparks.

The fascination with high-voltage work in the lab and outdoors in nature's magnificent displays was kindled in me back then and has grown over the years. Even at that early stage, I knew this was an area of research that was not well understood or explained, and I wondered whether such generators might ever have real, practical uses. Of course, today there are many industrial applications for electrostatics, such as removing dust from smoke stacks, paint spraying, and photocopying. But could an entirely new technology be developed for exploring the forces of nature and be as practical as was the application of steam power?

Modern electrostatics began about the year 1660, when Otto von Guericke built spinning sulphur spheres, which were charged by the friction of the hand. Considerable experimenting was done from the latter half of the 1700s to about 1900. By this time, increasing interest was developing in the applications of that other aspect of electrical science—electromagnetism. Michael Faraday, Nikola Tesla, and Thomas Edison were largely responsible for this development. In time, physics books would treat the study of electrostatics primarily as an entertaining novelty with no practical use.

In spite of the centuries of work, we still need much more understanding of the nature of electric "charge," of how electric forces act across space, and how electric potential energy is stored. At the very least, this would help to understand how thunderstorms develop.

This book is divided into two parts: the first part describes high-voltage generator design and construction with a brief mention of theory of operation. The second part details the basic instruments used with the generator and some of the areas wide open for pioneers. As the title implies, this book is not written for the couch potato who waits passively to be entertained, but rather for those who love to build, experiment, and investigate along original lines of research. I am especially interested in encouraging high school students and their teachers who want to do something really unusual for their science fair projects. A basic grasp of electrostatic principles is helpful on the theoretical side.

In this edition, I have simplified and improved the design and construction of my Wimshurst generator (with two counter-rotating discs). Realizing that many high school and university science labs already have Van de Graaff belt-type generators, I have devoted an entire chapter to Van de Graaffs, including principles of operation, construction, modifications for improvement, and the making of accessories. For both disc and belt generators, I have added notes on the general care, maintenance, and trouble-shooting peculiar to each type.

An intermediate level of skill in wood, metal, and plastics working, as given in high school industrial arts classes, is needed for generator construction. Very little math is required for the projects and experiments included in this book. I have provided an extensive bibliography for those students needing background research sources.

In this expanded third edition, I have included a description of a large sectorless Wimshurst generator with 24-inch discs for use in university physics lecture demonstrations; this is a scaled-up version of my original design. I also provide a new and simpler method for making electroscopes, building large capacitors using 5-gallon plastic buckets, and two new designs for the electrophorus including a large size for lecture demonstrations. New information on unusual electrical discharges, revised material suppliers, and extensive bibliography bring this revised edition up to date.

In order to get as many readers involved as possible, I have designed my generators and accessories with the idea of reasonable cost and readily available materials foremost in thought. I provide sources for parts not locally available.

Dedication

For my family and friends

Acknowledgments

Special thanks to A. D. Moore, whose books got me interested in electrostatics; to Franklin B. Lee, who has provided affordable, high-quality electrical apparatus to the public over a period of many years; to Diana whose modeling photos were borrowed from the first and second editions; to Kristine for the considerable time she offered modeling photos for this third edition; to Loomis Laboratory of Physics at the University of Illinois, Urbana-Champaign campus, for the use of lab and lecture equipment contained in their extensive and unique electrostatics collection; to the Urbana Free Library for their speedy interlibrary loan service; and to Liz for her computer and typing skills.

Lightning/ electrocution safety

Although lightning feels no compulsion to obey Ohm's law and it remains unpredictable, a few commonsense rules will reduce the hazards.

- Stay away from large bodies of water and wide-open spaces.
- When indoors, stay away from windows and fireplaces. Never talk on the telephone during violent storms.
- Stay in the car if you are isolated. All-metal cars, trains, and airplanes are fairly safe. Fiberglass or composite-fiber bodies will require a special means for diverting a strike.

In the case of an apparent electrocution, the victim should be treated as anyone needing CPR would be. Should this fail to get results, all is not lost as is illustrated by this case, mentioned many years ago in a science journal. A workman was electrocuted when he contacted a high-voltage power line. After all efforts to revive him failed, a most novel idea was hatched. As the victim lay on his back, one person removed his shoes and socks, and lifted up his legs, and held them together. A second person, with a large flat paddle, administered sharp blows to the soles of the workman's feet. Shortly, he was aroused and recovered fully—sort of like spanking the newborn baby, I suppose! I will leave it to experts to explain how this peculiar method of resuscitation worked.

Remember, the victim of an apparent electrocution might appear clinically dead, yet still not be beyond hope.

For updated information, contact the National Lightning Safety Institute, 891 N. Hoover Ave., Louisville, CO 80027.

1
PART

Design and construction

In this section of the book, I will describe various types of electrostatic generators and what comprises efficient high-voltage generator design. Step-by-step instructions on building my own 12-inch influence machine are given in Chap. 4. Also included are unusual generator designs, Van de Graaff generators, and briefly, theories of generator operation.

1
CHAPTER

Types of electrostatic generators

There are presently two main classes of electrostatic generators. The first type charges by *frictional slippage*, or *impact*. This means there is direct physical contact between two different material surfaces. Frictional charging in an earlier time was called *triboelectricity*, *tribo* being the Greek term meaning "to rub." Otto von Guericke's spinning sulfur sphere (1663), which rubbed against the hand, was the early form. Later in 1768, Mr. Jesse Ramsden and Mr. Jan Ingenhousz developed glass disc generators that rubbed against a leather pad coated with metallic powder instead of the hand. The frictional generator using glass discs reached a high level of development in 1856 with Karl Winter's design. For a good description of frictional generators, see the book *Early Electrical Machines* by Bern Dibner, listed in the Bibliography.

The second class of machines are called *electrostatic induction generators*. The word *induction*, according to the *Oxford English Dictionary*, in this context means to bring about an electrified state in a body by *proximity*—closeness without contact—of another electrified body. The induction generator was originally called an *influence machine*; the change in names occurred gradually during the years 1890 to 1920. I prefer the word *influence* because it gives a clearer mental picture. Oxford defines influence as "the action or inflow of immaterial things, the operation or infusion of which is unseen or insensible"—that is, only the *effects* are visible. The word *induction* draws a blank in the mind because it avoids all mention of the operating causes. I will discuss this theoretical subject later in detail. It is at the heart of a better understanding of just how electrification occurs, and especially the nature and properties of space itself. But first, I will describe the generators, which are our tools, as well as the novelties of their design and construction.

2
CHAPTER

Elements of good design

After two years of intensive research on both frictional and and influence generators dating from the late 1700s through the 1920s, I've concluded that one of the best machines, as far as general design is concerned, was the one that appeared circa 1856 in Vienna, Austria. The inventor, electrical genius Karl Winter, put together the best of his design innovations. So popular and effective were his machines that they were still in use by electrotherapeutics practitioners up to 1930! After having tried out several different generator designs, I find myself inevitably returning to his basic design elements, which can be applied to any high-voltage disc or drum generator.

Shown in Fig. 2-1, Winter's machine was greatly popularized throughout Europe after its introduction into Edinburgh, Scotland, by Dr. Robert M. Ferguson, who also mentioned it in his books on natural philosophy (physics). A rare description appears in the *English Mechanic*, which I have rearranged.

Winter's electrical machine[1]

In number 27, we expressed our determination to close our columns for the present to the subject of electricity, but the flood of letters which have come to hand has induced us to keep open for a week or two longer. To satisfy a crowd of correspondents, who seem altogether to have forgotten that what they have asked for has been published at length, and to enlighten several on a few new points of interest, we give an illustration and description of the machine invented by Herr Carl Winter, of Vienna.

It was first introduced into this country by Dr. R.M. Ferguson, of Edinburgh, in 1857. The Doctor had seen one in Vienna and different cities in Germany before he had one made. The largest he saw was in the Vienna Polytechnic School; it had a plate 5 feet in diameter which gave off sparks 4 feet long. However, a very good size is a 2-foot plate. At first the pillars

1 From the *English Mechanic*, October 20, 1865

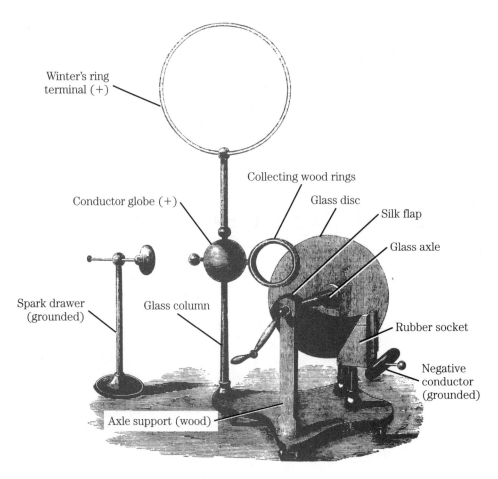

Winter's ring terminal (+)

Collecting wood rings

Conductor globe (+)

Glass disc

Silk flap

Glass axle

Spark drawer (grounded)

Glass column

Rubber socket

Negative conductor (grounded)

Axle support (wood)

2-1 Karl Winter's machine (1856). Labels added. *English Mechanic*, 1865

were made of glass, but Mr. Varley, of London, made his large-sized one of vulcanite, and found the insulation improved.

The machine is essentially a plate-friction machine, and with a 2-foot plate will give 11 inch sparks. If made in the ordinary way, such a machine would be considered very successful if it gave, in the most favourable circumstances, a spark of 5 or at most 6 inches. All who have had anything to do with electrical machines know how extremely difficult it is, even with the requisite means of drying, to keep them in a state favourable to ready action for any length of time. Mr. Winter left his machine in a room for months where, during that time, there had been no fire, yet by the first turn of the plate, it gave sparks 10 inches long. The cause of this efficiency may principally be attributed to the rubbers he employed, the perfection of insulation, and to the contrivance for lengthening the spark. The accompanying illustration represents one of these instruments.

It will be seen that the plate is fixed into an axle which revolves in two upright supports. One of these, in which the shorter wooden end of the axle revolves is made of glass, and the other, in which the longer glass end of the axle revolves, is of wood. Thus the electricity formed upon the plate cannot reach the ground on either side, for on the one side the insulating glass pillar, and on the other the insulating glass axle, prevent it.

The friction, as usual, is caused by pressing on the plate of glass a flat surface of leather covered with an amalgam of mercury, zinc, and tin, which is put on with the aid of a little grease. The frame standing on the low glass support to the right of the figure is the wooden rubber socket, into the notches of which fit two flat pieces of wood covered in front or on the side next to the plate with leather, and a very little stuffing. These leather pads are provided on the other side with springs, which, acting against the frame, keep the front surface uniformly pressed against the plate. There is only one pair of rubbers, not two, as in ordinary machines, and this enables them to be placed at a greater distance from the prime conductor of the machine.

The brass ball on the tall glass support to the left is the prime conductor. For more perfect insulation this ball is fitted on to the support by means of a trumpet-shaped opening made in it, thereby preventing the dispersion of electricity that would arise from the sharp edge of a hole exactly large enough for the rod. There are three other openings in this ball, one on each side, and one at the top. The two small rings which are seen projecting upon the plate fit into one of these by means of a T-shaped piece of brass. They are made of wood, and have a groove cut in them on the side turned towards the plate, into which a row of fine pin points is fixed for collecting the electricity formed upon it.

These points are connected with the prime conductor by means of a strip of tin-foil which lines the bottom of the groove. Two wings of oil silk, attached to the rubbers, stretch between them and these rings, so as to prevent the electricity from dispersing itself before reaching them. The opening in the top of the ball is made to receive the stalk of the wooden ring, which is seen surmounting it, and which forms the most peculiar feature of the instrument. The function performed by this remarkable appendage is to lengthen the sparks given by the machine. It is a wooden ring, and might be from 32 to 36 inches in diameter, with a thin iron wire inside to form a core, which also passes down the wooden stalk, and is in metallic connection with the brass ball or prime conductor.

This ring can be removed at pleasure; when removed, it is an ordinary plate machine of the best construction. The sparks then are straight and thick, and not more than two or three inches long, whenever the ring is put on they are at once lengthened to 10 or 12 inches. It would seem that the electricity accumulates with great intensity on the thin wire which forms the core above noted, the wooden covering preventing dispersion.

Dr. Ferguson tried different rings; he had one of polished iron wire, and got 6 in. sparks; covered with gutta percha he obtained 12 in. sparks; puncturing the covering the length of spark fell to 6 in. Other kinds of rings were tried, but none proved so good as that described. This is probably owing to its semi-conducting power, or superior elasticity, or in that it cannot be permanently punctured like gutta percha.

To the left of the figure is the spark drawer, for receiving the sparks from the machine. The length of the spark is the test of the construction of a machine, and it would appear that in this respect Winter's holds first rank. Indeed, it is something quite novel in the history of electrical apparatus; concerning which, Mr. W. Hart of No. 7, North College Street, Edinburgh, has expressed his willingness to answer correspondents by letter.

Of special interest is Winter's extensive use of wood, a semiconductor, for high-voltage charge collection. Further details were given in *Physical Technics* by Dr. J. Frick (see Bibliography).

The collecting apparatus consists of two thick, polished wooden rings, from 2 to 5 centimeters thick with an external diameter of 13 centimeters. A groove is cut on the inside of each ring—facing the glass disc. This groove is lined with tin-foil and set with a thick row of fine pin points, which reach only to the surface of the ring. The elements of good generator design used by Karl Winter include the use of large, smooth surfaces of polished wood for collecting electricity from the disc.

Many electricians in later years forgot this detail and made small wire collectors, which easily leak charges at high potential. Evidently, these later inventors had not studied or been aware of Winter's designs, the result being that the voltage or charge potential produced in those generators was inferior. By surrounding metal with wood, a gradual change of resistance from the conductor outward through wood (semiconductor) to the surrounding air (insulator) results. This absence of an abrupt change further reduces high-voltage leakage.

I have no doubt that several of the early frictional generators could outperform many of the later influence machines. In Winter's design, spark length was four-fifths the diameter of the glass disc, which is very good! The current output, in those days, was difficult to measure.

Much of the history of these early electrical devices was lost because the letters describing them have since turned to dust. Karl Winter's ingenious design is one of the best from that era.

3
CHAPTER

James Wimshurst's influence machine

The basic influence (induction) generator had its beginnings with John Canton's *theory of electrification without touching* from about 1753. He visualized an electrical *atmosphere*, a medium that surrounds electric bodies and acts across space.

Later in 1787, Abraham Bennet described a "doubler" generator, which via influence continually increased very small charges until they built up to measurable values. Throughout the 1800s, new designs appeared; their number began to accelerate about 1860.

It is not possible to describe all the designs, but the most popular were Varley's machine (1860), Toepler's machine (1865), Holtz's machine (1865), Leyser's machine (1873), and Voss' machine (1880). Of these, the Holtz generator would continue to be used for many years in electrotherapeutics and was able to outperform the Wimshurst generator in good weather.

But these earlier designs suffered from either difficulty in starting in bad weather or from polarity reversal. Reversal of polarity between the two output terminals could occur spontaneously, so constant reliable output was hampered. In 1878, James Wimshurst of England set about to remedy these problems and improve on the Holtz design. In 1883, the basic Wimshurst machine began to appear in the science journals. His original design is shown in Fig. 3-1.

As can be seen in this figure, no provision was made for storing charges. In 1882, charge storage was added by using two Leyden jars, the first capacitors. This method increased the efficiency and increased the spark length between the discharge terminal balls.

The Wimshurst generator proved to be quite reliable in starting during the damp weather in England because of the metallic sectors on the two discs. A single turn of the handle would cause the two glass plates to literally bristle with electricity. In addition, Wimshurst's design was not subject to polarity reversal like other machines.

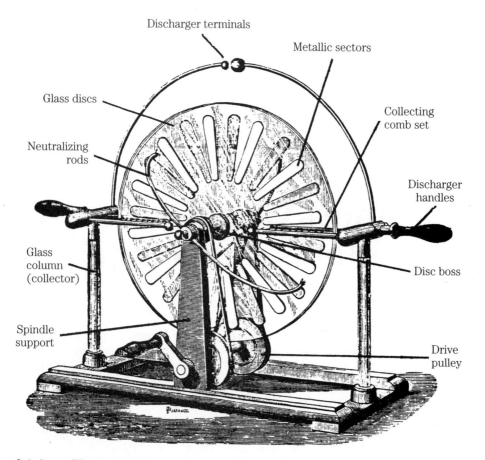

3-1 James Wimshurst's original design. *Electrical Influence Machines*, John Gray, 1903

These two desirable features resulted in far-reaching popularity for the Wimshurst machine in Europe. Many articles appeared on variations of his basic concepts.

The largest Wimshurst was built in 1885 and had two 7-foot glass discs ⅜ inch thick, each weighing 280 pounds! (See Fig. 3-2.) When operating, this machine gave a torrent of hot sparks and electric flames up to 22 inches long.

I did manage to track this device down. As of this writing, it is located in a glass case on the second floor balcony of the Science and Industry Museum in Chicago, Illinois. I found it inoperable and in need of major repairs, but still impressive to see. Perhaps the museum staff will see fit to put it into operation for the benefit of scientific inquirers.

Another variation of the design enclosed the machine in an airtight case with chemical drying agents. Dry air helped boost voltage output and made starting easier. (See Fig. 3-3.) This commercial design was used by electrotherapeutic practitioners. The charge collector has been enlarged in size but is still entirely made of metal. By using multiple glass plates on a single axle, the current could be increased just as one would join batteries in a parallel circuit. (The voltage remains the same.)

3-2 The largest Wimshurst machine (1885). *Engineering*, vol. 39, 1885

The bottom circuit for the Leyden jar condensers use a wire to join and ground the four outer coatings of the four condensers. *Ground* here means through the woodwork, which is usually sufficient. The inner coatings of each pair of condensers is joined to its respective collecting comb. The Leyden jars should be crystal, crown, or "Pyrex-type" glass so that high-voltage leakage is reduced. The collecting points can be steel dressmaker's pins; the gap between pinpoints and disc surface should be greater than $\frac{1}{16}$ inch, but not more than $\frac{1}{8}$ inch.

Notice that the discharging balls are different sizes. This geometry results in a greater spark length. For an 18-inch-diameter disc machine, Mr. Wimshurst preferred

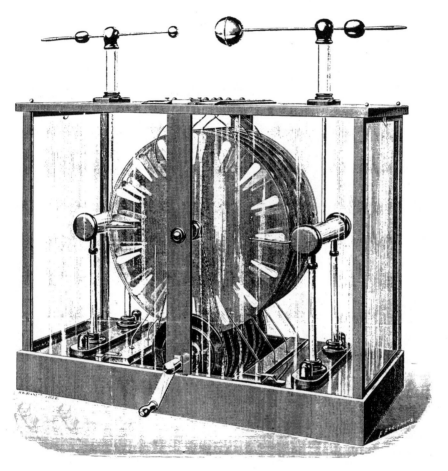

3-3 A multiplate Wimshurst (1888). *Engineering, vol. 45, 1888*

a small ball ½ inch in diameter and a larger ball of 1½ inches in diameter. These two balls should be interchangeable, rather than permanently soldered on, and should depend on terminal polarity for placement.

The neutralizer brushes must make contact with the disc's metallic sectors simultaneously across the disc diameter. For this reason, only an even number of sectors can be used. Bend the arms slightly to make these contacts possible.

Brushes for the neutralizers should be springy, since they make light contact with the metallic sectors on the discs. Phosphor bronze or steel music wire 0.002 to 0.005 inch in diameter are serviceable for this purpose, but they should not be so stiff that they wear away the sectors. The metallic sectors can be cut from brass or steel shimstock or adhesive-back stainless steel foil, 0.001 to 0.003 inch thick. These sectors should be cut out with sharp paper scissors, not with metal snips; snips leave a serrated edge that causes leakage. I recommend attaching sectors with epoxy glue since shellac varnish is not a good adhesive in damp weather. The sectors should be attached after the discs are varnished.

The method adopted for varnishing the glass discs, to make the surface resistant to moisture films is as follows.

Wimshurst machine[1]

The following is the best method for varnishing the glass discs: first mix your varnish (shellac and methylated spirits) in a bottle having a large neck; the brush ought to be more than 1 in. in diameter, and its handle should be fixed in the bung of the bottle; this arrangement keeps the varnish and the brush free from dust. Next make up some simple sort of turntable, on which you place the disc while it is being varnished. The next step is to wash the discs and dry them; then warm each disc to about the temperature of blood heat, place one on the turntable, and with the left hand set it in motion. Take the brush full of varnish from the bottle with the right hand and bring it on to the disc a little from the edge; move the brush slowly to the edge and then back, so as to finish and to lift the brush off at the centre; let the disc remain a few minutes for the varnish to set, and then finish drying in a warm place. It will then be seen the coating is so even as to be almost unnoticed. If the brush be lifted up from the disc while the varnishing is being done, the surface will be covered with air bubbles, and look unsightly. Done in this way the varnish lasts for years.

Warming the discs drives off moisture; the room should be dust-free and dry, of course. *Methylated spirits* is also called *denatured alcohol solvent*, and the shellac is in the form of refined flakes. If you add one or two drops of castor oil per pint of varnish, there will be less tendency for the varnish surface to crack and dry out.

The usual means for attaching the glass discs to the wood bosses is as follows:

Cut out a thin leather washer, about $1/32$ to $1/16$ inch with an outer diameter equal to the diameter of the boss. The hole in the washer should be a bit larger than the shaft through the discs. Cover one side of the washer with epoxy glue, or any cement for repairing glass, and press the tacky side onto the boss end. Coat the top side of the washer, and let it stand until the glue is tacky. Press the boss onto the glass disc, carefully centering, so that the boss hole is concentric with the disc diameter. Apply weight on top until the glue is dry, usually overnight.

Mr. Wimshurst also used thin leather washers cemented with bicycle tire cement. Always use washers because they cushion the discs and prevent cracking. Finally, the discs are spaced apart from each other, about $1/32$ inch by using metal, fiber, or leather washers as spacers—the discs must not rub or touch at any point.

The traditional Wimshurst machine was hand-powered. Although such machines can be motor-powered for convenience, as my generator is, the speed of the glass discs must be kept low to avoid breakage. There is a certain fascination with hand-powered versions because they produce miniature lightning bolts without any attached power cord—seemingly drawing electricity out of thin air.

---

1 From the *English Mechanic*, 1889.

Modified Wimshurst machines

In spite of the glowing reports in English science journals about the performance of Wimshurst's generators, there were detractors who pointed out defects in his designs. Some of these inventors put forward their own innovations, which proved to be more efficient.

Under the best of conditions, the traditional Wimshurst could develop a spark length equal to the radius of the disc, but this was rare for most homemade units. The current or quantity of charge was just as important as spark length. The usual method for comparing two designs of the same size and speed was to see how fast each could charge a Leyden jar to a fixed voltage. Results were not very accurate and this area of research needs to be explored. It should be clear that influence machines basically have their voltages determined by disc diameter, so doubling the diameter approximately doubles the voltage. Current output also increases with disc rotational speed, and of course with added pairs of discs on the same axle.

Methods for improving efficiency will be mentioned later. During the 1890s and onward, modified Wimshurst machines appeared. One of the important changes was to remove all sectors on the discs. These generators then became known as *sectorless Wimshurst machines*. Most notable were the inventors Mr. Picolet (1892) and Mr. Bonetti (1893). Figure 3-4 describes Picolet's design.

3-4 A sectorless Wimshurst (1894). *Scientific American, 1894*

Sectors were found to be a cause of serious leakage, and their only advantage was to make the machine self-starting. I find that having to "charge" the sectorless generator is simple and is a good safety feature. In addition, the plates can easily be kept clean. As shown in Picolet's generator (*Scientific American*, May 1894), the neutralizer brushes are long with several points, but these points do not need to touch the disc.

Another solution to the metallic sector leakage problem was put forward in two patents by Mr. Wommelsdorf (U.S. patents 882,508 March 1908 and 1,071,196 August 1913). He sandwiched his sectors, or *carriers*, within two thin discs. Electric contact to these carriers was through the compound disc rim only. Even though his generator was not a true Wimshurst, the technique would still apply. Wommelsdorf claimed a large improvement in current output and a spark length of two-thirds the disc diameter—quite good! In addition, his unit was not sensitive to weather changes.

A different modification was tested by the inventor Lemstrom in his work on electrification of crops (U.S. patents 634,467 October 1899 and 720,711 February 1903). He made his influence machines using counter-rotating drums, instead of discs, for compactness and improved current output. Being compact, Lemstrom's generator could be enclosed in a wood cabinet with chemically dried air. His generator could be run for long periods without maintenance and was not sensitive to weather changes.

In the last modification, I consider a Mr. Schaffers who put forth his theory of how Wimshurst machines work in July 1895 in vol. 35 of *The Electrician*. His main innovation was to alter the shape and placement of the collector combs.

Even though Wimshurst had experimented with different comb designs, the traditional shape was a simple U placed in the horizontal diameter position as seen in most photos. Schaffers skewed the comb arms by about 60 degrees from each other after finding the best places on the two discs to pick up charges. Current output was improved. Most importantly, Schaffers' discovery did illustrate the considerable lack of agreement on generator operation theories. Of course, a good theory would lead to improved efficiency, greater voltage and current output.

4
CHAPTER

The author's generators

The following is a detailed description of my 12-inch influence machine (Fig. 4-1), which is modified in several ways from James Wimshurst's original design. Please review Fig. 3-1 to familiarize yourself with the parts of the generator that are described in the construction notes. Read through that section entirely before attempting to build.

The design concept is very simple. The generator is motor-powered with each disc being driven directly by its own motor. This eliminates belt problems and by eliminating the hand crank, the experimenter frees his hands to do experiments.

As made, my generator with Leyden jars can produce sparks up to 10 inches in overall length, using 12-inch-diameter discs. This is 9/10 of the disc diameter, which is quite good! Both the Holtz and the Wommelsdorf generators had good sparking distances, normally two-thirds the disc diameter.

Special Note: Those who want to copy my results should duplicate both the dimensions and the materials I specify. Any departure can cause high-voltage leakage or reduced current output and spoil your enthusiasm. Once you have made a good unit with comparable results, then you can begin to experiment on the infinite changes possible in design.

The overall dimensions of the generator's wood stand were sized to fit the drive motors, which are 12-Vdc permanent magnet motors. I chose 12 Vdc because such motors are quiet, powerful, small, and easily varied in speed. If you live in a heavily populated area where the neighbors take a dim view of folks who make lightning bolts, you can drive to a remote site and power your Wimshurst through the vehicle's cigarette lighter outlet.

The stand and upright supports

The wooden parts of the machine should be made from well-seasoned lumber, free from knots. Aspen, poplar, and oak are commonly available and machine well.

The stand consists of two side rails measuring ¾ × 2 × 16 inches. Four scroll feet are cut from stock measuring ½ × 4 × 7½ inches (Figs. 4-2 and 4-3). Each pair of

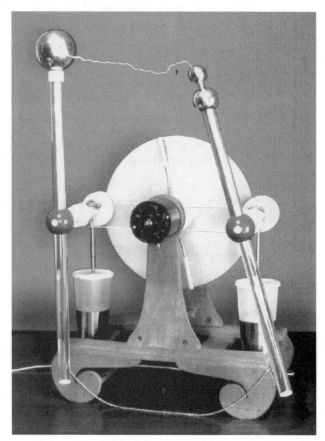

4-1 Author's sectorless Wimshurst generator with 12-inch discs.

scroll feet is joined together as a unit with three crossties. A total of two crossties measuring ¾ × 2 × 9 inches and four crossties ½ × 2 × 9 are needed. To fancy up the appearance of the stand, I chose a ⁵⁄₁₆-inch-radius "corner rounding" router bit (Fig. 4-4). I don't have a router, so I cut the mouldings using the highest speed on my drill press, taking several light cuts.

The two pairs of scroll feet are then jointed to the two side rails as a unit with four 1-inch-long drywall screws (Fig. 4-5). The two upright supports that hold the motors are each sandwiched together from ½-inch stock. The uprights are cut from four pieces of stock ½ × 5½ × 12½ (Fig. 4-6).

Refer to the drawing for positions of holes, grooves, and slots required in the uprights. Next, temporarily join each pair of uprights together using small ¾-inch brad nails and cut a tenon on the bottom and the mortise slot in the side rails (Fig. 4-7). The mortise and tenon joint secures the uprights to the base so they are perpendicular and parallel. When finally assembled, the two uprights are pulled down tight using furniture joint connectors and socket head bolts (Fig. 4-8). This arrangement allows uprights and motors to be removed easily for changing the machine's discs.

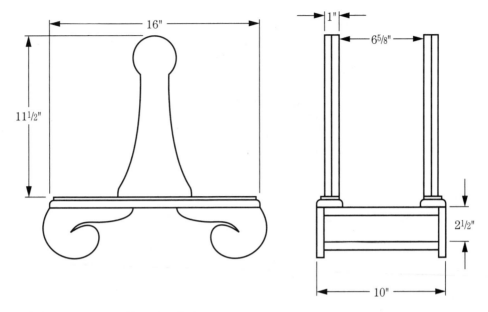

4-2 Generator stand, Victorian design.

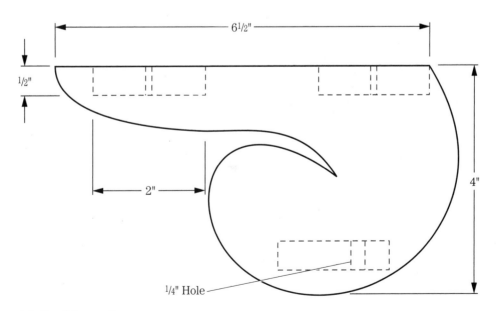

4-3 Scroll feet pattern.

4-4 Corner rounding router bit and molding.

4-5
Scroll feet and crossties, unit.

Motors and mounting

The matched pair of motors I used were Redmond brand 12-Vdc permanent magnet reversibles, 3300 rpm no-load speed; 1300 rpm and 1.3 amps with 60 oz-in. torque.

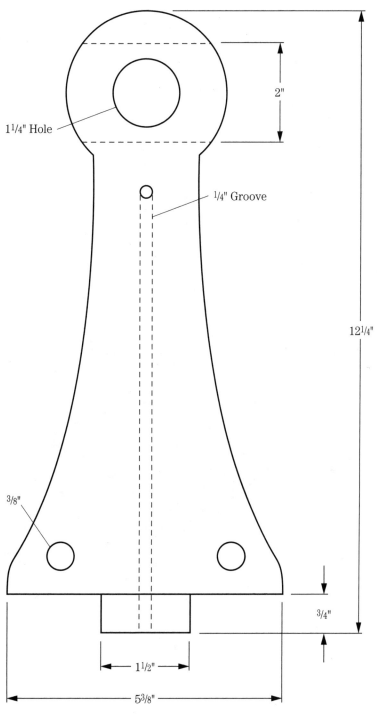

4-6 Upright support pattern.

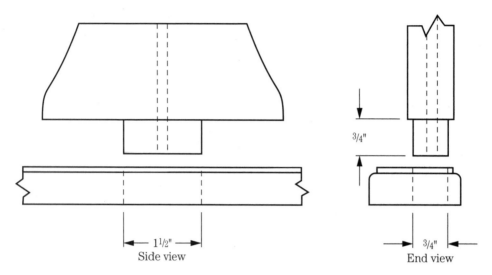

Side view   End view

4-7 Mortise and tenon joint, upright.

The shaft was ⁵⁄₁₆ inch diameter × 2½ inches long. Dimensions of the motor body (minus shaft) are 3¹⁄₁₆ inches diameter × 3¾ inches long. Choose motors with similar characteristics. If the supplier's motor shaft is short (less than 1 inch in length is typical) add "motor shaft extenders." These increase the shaft length by about 1¼ inches and are secured with a setscrew. This longer shaft length helps support the discs and prevent wobbling (Fig. 4-9).

4-8 Furniture connectors and socket head bolts.

4-9
Motor and shaft extender.

Each motor is centered in the hole at the top of its upright and either secured with the two existing motor stud bolts, or two holes are drilled and tapped into the face of the motor, which is then secured with #8-32 machine screws passing through the upright (Figs. 4-10 and 4-11). Because each upright is a two-piece sandwich with a groove running down the middle inside, the two wires from the motor pass into and snake down through the upright and the side rail as well (Fig. 4-12). Note that the two motor mount screws are offset from the central raceway so as to not obstruct the wires.

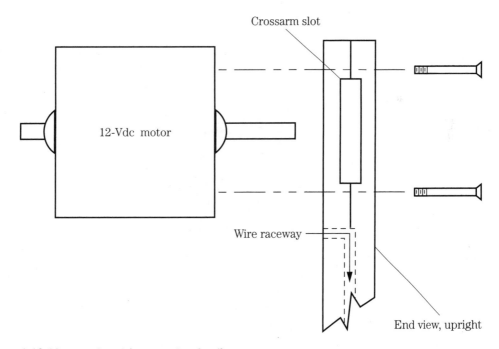

Crossarm slot

12-Vdc motor

Wire raceway

End view, upright

4-10 Motor and upright mounting detail.

4-11 Motor mounted to upright.

The discs

The two discs are cut from plastic sheet having good dielectric properties and dimensional stability. For this particular machine, the 12-inch-diameter discs are ⅛ inch thick to ³⁄₁₆ inch thick. The only exception to this rule would be if you experiment with 33 rpm vinyl records. Test any sheet by cleaning with denatured alcohol, then rub briskly with a clean cotton cloth to see if it holds a charge well. Place a

4-12 Motor wires concealed in upright.

12-inch steel rule across the surface with the sheet held vertically to be sure it is flat, not warped.

Although acrylic plastic may be used, I do not prefer it because it is seldom flat and is unstable at very warm temperatures. An especially good disc material is G-10 epoxy/glass, which is a type of phenolic laminated sheet. It has a high dielectric constant, which increases current output of the generator, and has low moisture absorption so that charge doesn't leak off the discs easily. Consult with your plastics supplier for physical properties of other kinds of plastic sheet suitable for discs. G-10, being abrasive, is best cut out with a sabre saw or jigsaw using a high-speed steel metal cutting blade having about 18 teeth per inch. Accurately locate a center hole sized to fit your motor shaft (mine is ⁵⁄₁₆ inch diameter). The edge of each disc is trued for balancing later.

Bosses and bushings for discs

These parts must be machined accurately on the metal lathe because the discs must spin true to get reliable generator output. The bosses are the round projections that attach directly to the discs (Fig. 4-13). The two bosses are ¾ inch thick × 3 inches diameter and are cut out of a plate of gray PVC (polyvinylchloride), or delrin. These plastics machine easily and are dimensionally stable. Originally the bosses were wood (see Fig. 3-1), but this can cause leakage across the discs at high voltage. Six equidistant holes are drilled and tapped for size #10-24 thread in each boss face. A ⁵⁄₈-inch hole (0.625") is bored through the center of the boss.

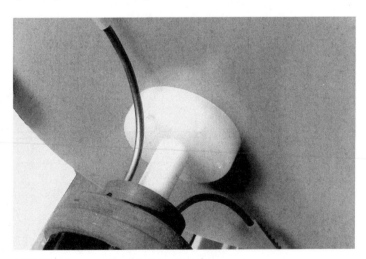

4-13 Disc boss and bushing, mounted.

Plastic bushings of the same material are cut from ¾-inch rod and each is 2¾ inches long. A hole is bored through the center of each bushing sized accurately for your motor shaft. In my case it was 0.312 inch for the ⁵⁄₁₆-inch shaft. On the motor side of each bushing I slit it down 1¼ inch with a hacksaw so that it clamps onto the shaft with a sliding fit. Re-bore bushing after slitting if necessary to remove the rough edges of plastic. If you are using motor shaft extensions, re-bore the motor side of the bushing to accommodate the enlarged diameter of the extension's shoulder (where the setscrew is located).

To properly fit each bushing to its boss, I turn a shoulder ⅝ inch diameter (0.626") and ¾ inch long and press the bushing into the boss with a vise. I next mount the boss and bushing unit on a 4-inch section of drill rod having the same diameter as the motor shaft and clamp the split end of bushing with a hose clamp. Finally, I face off the boss surface next to the disc. By this method, I establish a flat surface that is perpendicular to the shaft so that the disc will not wobble and spoil generator output.

I accurately mark the six corresponding holes in the disc centered on its boss and drill and countersink holes for ³⁄₁₆ inch. Each disc is then attached to its mating boss with #10-24 flathead nylon screws ½ to ¾ inch long. (Nylon screws also reduce charge leakage.) Note that screw heads are flush with the discs' inner surface, since the discs spin in opposite directions.

The boss and disc unit is now ready for balancing. A simple jig (Fig. 4-14) is made to allow for finishing the disc edge so it is truly concentric. The "pivot sanding" jig is a ¾-inch piece of plywood with the before-mentioned drill rod pressed into it to act as pivot for each disc. Clamp this platform down on the sander's table, put on the disc and boss unit, and adjust until the disc edge contacts the sanding disc or belt. Then slowly turn the disc to give a perfect circular edge. Take several light cuts to avoid distorting the work. If you have done this procedure with care, the discs will be balanced at high-speed settings.

If needed, plastic locking collars (Fig. 4-15) can be installed over the split end of the bushing to secure it to the motor shaft. These collars may use steel #10-24 setscrews.

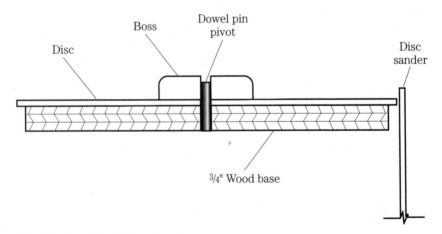

4-14 Pivot sanding jig for disc edge.

Neutralizer blades and mountings

Figures 4-16 and 4-17 show a neutralizer blade unit. Two of these units are needed.

Into the inside half of each upright support (opposite the motors), I attach a metal ring, which projects out ½ inch (Fig. 4-18). Onto this ring I can mount a

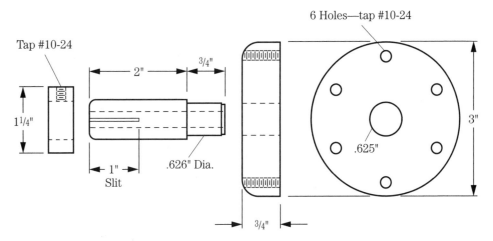

4-15 Locking collar, bushing, and boss.

neutralizer blade unit so it can be pivoted or moved toward or away from its disc. Each ring can be cut from a single sweat copper coupling for 1-inch pipe. The plastic neutralizer blade collars are each turned from PVC or delrin plate and are ¾ inch thick × 2 inches diameter. A 1¼-inch hole is bored through to make a ring, and into this ring face I drill two ³⁄₁₆-inch holes diametrically opposite each other.

4-16
Neutralizer blade unit.

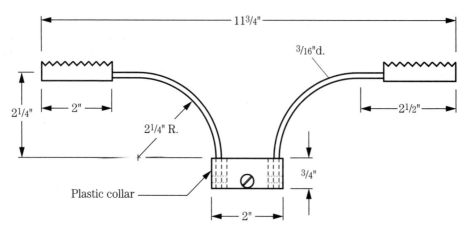

4-17 Neutralizer unit detail.

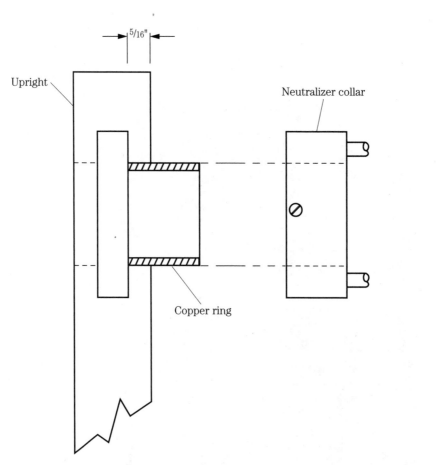

4-18 Copper ring, neutralizer mounting.

A ³⁄₁₆-inch brass rod with rounded ends is bent as shown (actual length is 6¾ inch) and two such rods form the arms for the blades. These are pressed into the face of the plastic ring and project up from the ring surface 2¼ inches. Note that the outer 3 inches of each rod is parallel to the disc. The outer ends of these two rods are 11¾ inches apart. To be sure that the two rods electrically conduct with each other as a unit, I slipped in a U-shaped piece of shimstock 0.002-×-⅛-×-1-inch overall length into the thru-holes seen in the plastic ring. These brushes straddle the brass rods, projecting inside the ring enough to contact the copper mounting rings (Fig. 4-19).

Each neutralizer blade, four required for the generator, is cut from 0.002-inch brass shimstock ⅞ inch wide × 2 inches long. The serrated edges (Fig. 4-20) are cut with dressmaker's pinking shears so that the points are in a straight line. The shimstock blade is first curled over a ³⁄₁₆-inch rod to form a cylinder. Blade holders are

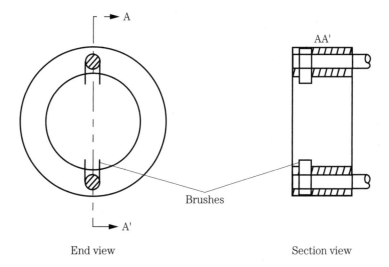

End view   Section view

4-19 Neutralizer, plastic collar, and contact brushes.

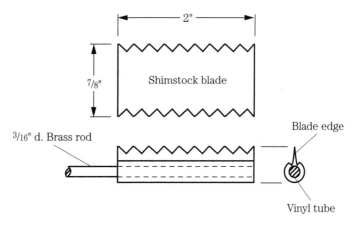

4-20 Neutralizer blade detail.

made from sections of polyethylene tubing ⁵⁄₁₆ inch o.d. × ¼ inch i.d. × 2¼ inches long. Slit each tube lengthwise with a utility knife and insert the rolled-up brass blade so that the serrated edges stick out of the slit about ³⁄₁₆ inch.

Slide each of the four blades onto the brass rod ends of the neutralizer units. Outer edges of each blade pair should be about 11⅞ inches apart. In each plastic ring I drill and tap a hole as shown for #10-24 and insert a 1-inch-long nylon screw, which secures the neutralizer to its copper ring mounting on the upright support.

Install neutralizers onto copper rings, taking note of blade positions and direction of disc rotation (Fig. 4-21). This is essential for your generator to begin charging. Check blade points to be sure they are parallel to the disc surface with the air gap being about ¹⁄₁₆ inch. These blades should not touch the disc at any point.

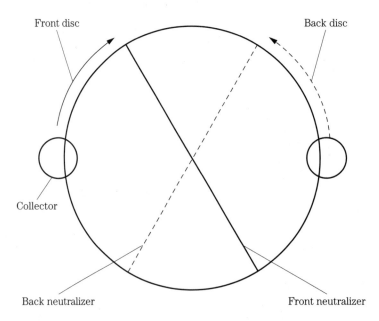

4-21 Neutralizers and discs positioning.

Power supply and wiring hookups

The best method for powering the 12-Vdc motors is to use a 12-Vdc battery charger rated for at least 4 to 5 amps capacity.

The two motors are wired together in parallel, the four wires being soldered to the four pins on a mic jack (Radio Shack part #274-002A). A 4-pin mic plug (Radio Shack part #274-001A) is joined to a 4-wire conductor (22-gauge round telephone hookup wire) several feet in length. The jack can be attached with an angle aluminum bracket to the bottom of a crosstie between the scroll feet (Fig. 4-22).

One pair of conductors leads to one clip of the battery charger and the other passes through a speed-controlling rheostat, such as Ohmite 8-ohm, 100-watt rheostat. This size gives good control for the specific motors I used.

4-22 4-pin jack and plug attachment.

You may otherwise make up your own power supply with a step-down transformer for 110-Vac primary to 12 Vac at 4 amps for the secondary output. The two secondary wires are joined to a full-wave bridge rectifier (Radio Shack part #276-1185), which is rated for 25 amps at 50 PIV. Figures 4-23 and 4-24 illustrate alternate wiring hookups.

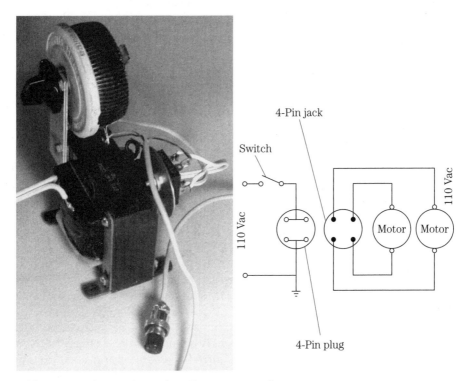

4-23 Homemade transformer/rectifier power supply.

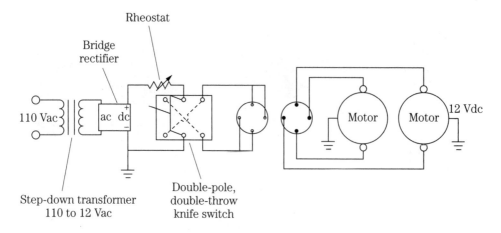

4-24 Motor circuit diagram with reversing switch.

If you are using 110-Vac motors, speed control may not be possible. Such motors should be rated for about $\frac{1}{10}$ horsepower, be a matched pair, and be quiet in operation. There are a number of fan motors with long shafts. If they are too long, shorten them with a tungsten carbide hacksaw blade (shafts are hardened).

At this point in construction, the generator will charge and run. Check neutralizer angle and gap setting and space the two discs apart $\frac{1}{32}$ to $\frac{1}{16}$ inch. No parts should rub while running. You can use your fingers as collectors along the horizontal line passing through the disc diameter.

Charge collector support system

Note that the front upright support, which is a sandwich of $\frac{1}{2}$-inch pieces, has a horizontal slot at the top so that a plastic bar can be clamped in place (Fig. 4-25). This collector support bar can be made of PVC, or if you want it clear, acrylic. It measures $\frac{3}{8} \times 2 \times 15\frac{1}{2}$ inches. Two $1\frac{5}{16}$-inch-diameter holes, spaced $13\frac{1}{2}$ inches apart, and one $1\frac{1}{4}$-inch-diameter center hole (to accommodate the copper ring neutralizer support) are needed.

The aluminum collector tubes (Fig. 4-26) are held by standard PVC 1-inch threaded plumbing caps with a hole bored for $\frac{7}{8}$ inch and a standard PVC 1-inch to $\frac{3}{4}$-inch reducer bushing. I turn down the outer threads on the bushing slightly and deepen those thread grooves so that the cap and bushing screw down on the plastic support bar. I re-bore the PVC bushing to $\frac{7}{8}$ inch for a snug sliding fit onto a piece of $\frac{7}{8}$-inch o.d. aluminum tube.

I then cut four slots along the bushing threads (Fig. 4-27) so that when tightened, it locks the tube in position in the bar. Two aluminum collector tubes $\frac{7}{8}$ inch o.d. × 8 inches long are cut. These are de-burred and carefully polished. All metal

4-25 Plastic bar mount in front upright.

parts of the collector system must be free from sharp edges and well polished to prevent charge leakage. Into one end of each tube I insert one ⅞-inch cork and #10-24 tee nut (Fig. 4-28) and recess this ⅛ inch deep. This can be set with epoxy glue.

4-26 Aluminum collector tube mounted.

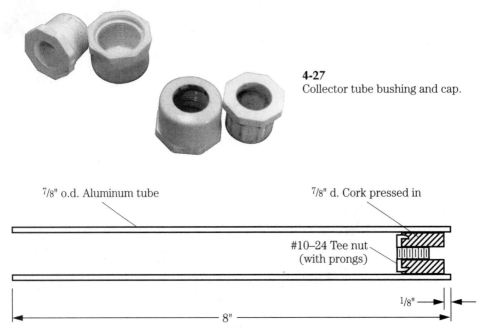

4-27
Collector tube bushing and cap.

⁷/8" o.d. Aluminum tube

⁷/8" d. Cork pressed in

#10–24 Tee nut
(with prongs)

1/8"

8"

4-28 Collector tube, internal detail.

Charge collector combs

The collector combs are housed within a cupped pair of standard 2-inch PVC pipe caps. Note the position of holes and slots. The discs will pass into the slotted area (Fig. 4-29). The ⅞-inch hole in the caps is cut to give a snug sliding fit over the aluminum tubes. The outer cap fastens to the internal tee nut with a nylon binding screw #10-24 × 1 inch long.

I line each of the four caps (Fig. 4-30) with a strip of shimstock 0.002 × ³/₁₆ × 3 inches, serrated on one edge which faces the disc. These are superglued so that the row of points is flush with the cap's slot. Strips of adhesive aluminum tape join the shimstock along the inside of the cap to its aluminum tube to give electrical contact. You might fashion caps from wood as Karl Winter did for his frictional generator. Sand these smooth and apply three coats of glossy polyurethane varnish.

4-29
Collectors made from PVC pipe caps.

4-30
Shimstock combs inside PVC caps.

Discharge handles and terminals

I prefer to have the negative polarity handle longer than the positive handle so that terminals line up better for longer sparks. The aluminum tube for the first handle is ⅞ inch o.d. × ¾ inch i.d. × 11½ inches long, and the positive handle is the same diameter but 10 inches long. The two insulated rods are clear acrylic ⅞ inch diameter × 12 inches long. I cut a shoulder 1 inch long and ¾ inch diameter and these will be press-fit into the aluminum tubes which have been de-burred and well polished (Figs. 4-31 and 4-32).

Into the bottom end of each acrylic handle, I cut a shallow depression, either conical or concave. I paint this depression with three coats of luminous paint. In the dark, this makes it easy to adjust handles. You should coat the rheostat speed control knob as well.

4-31
Discharge handles, aluminum and acrylic.

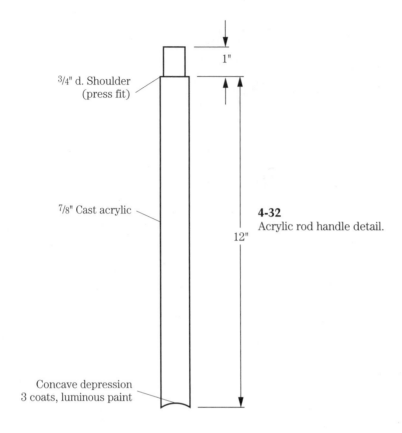

3/4" d. Shoulder
(press fit)

1"

7/8" Cast acrylic

12"

Concave depression
3 coats, luminous paint

4-32
Acrylic rod handle detail.

The two handles slide into sockets fashioned from 2-inch-diameter wood balls (Fig. 4-33). The thru-holes are ⅞ inch diameter and the front of each ball is threaded for a ¼-inch 20 nylon binding screw ⅝ inch long. To join the two balls to the ends of the aluminum tubes, I bore a hole 1 inch diameter and ⅜ inch deep and insert the plastic bushing and cement with 2-part epoxy glue (Figs. 4-34 and 4-35). Note that a ¼-inch 20 nylon binding screw for tightening handles is provided in each ball. The ball sockets are friction-fit to the ⅞-inch aluminum tubes.

Terminal caps are cut from PVC or delrin and use steel washers and ¼-inch-20 studbolts to join terminal balls to aluminum tubes (Fig. 4-36). The caps slip on with a light friction fit.

4-33
Two-inch wood ball handle socket.

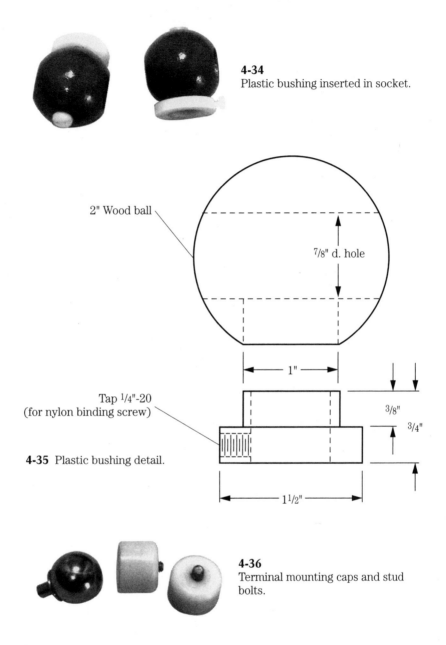

4-34
Plastic bushing inserted in socket.

2" Wood ball

7/8" d. hole

1"

Tap 1/4"-20
(for nylon binding screw)

3/8"

3/4"

4-35 Plastic bushing detail.

1 1/2"

4-36
Terminal mounting caps and stud bolts.

Two of the balls are fashioned from a pair of Chinese exercise balls (Fig. 4-37). The negative ball is 1¾ inches diameter with a hole tapped for ¼-inch-20 threads. When drilling these use sheet rubber or wood strips inside vise jaws to prevent damage to terminals. Tap out excess iron oxide inside the ball when the hole breaks through.

The double positive terminal uses the same size ball, which is also bored with a 1-inch hole and a ¼-inch hole on the opposite side (Fig. 4-38). The smaller ball on top

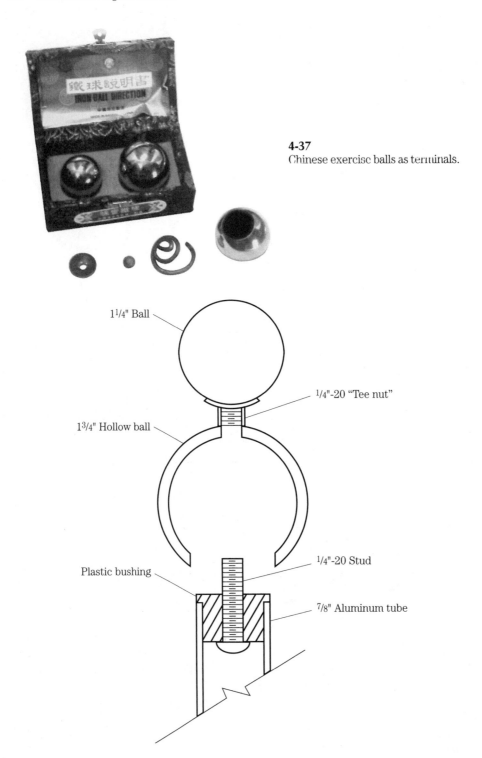

4-37
Chinese exercise balls as terminals.

1¹/₄" Ball

¹/₄"-20 "Tee nut"

1³/₄" Hollow ball

Plastic bushing

¹/₄"-20 Stud

7/8" Aluminum tube

4-38 Positive polarity double ball terminal.

is a 1¼-inch steel ball bearing. This is soldered onto a "tee nut" (available at the nuts and bolts section of your hardware store). The tee nuts are deformed by placing the ball bearing against the flange of the tee nut with a ¾-inch 10 hex nut as a backup support (Fig. 4-39).

When these balls are pressed in a vise, the flange assumes a cup shape which fits the ball. Clean and tin this flange and ball separately with silver solder, then turn the tee nut with the flange up in the vise. Place the ball on top and gently heat only the ball. Do not overheat; the ball should turn slightly purple and will nestle into the flange cup. Cool with dripping water, then clean off flux. Remove excess solder and buff to a high polish (Fig. 4-40). Remember that all metal terminal parts must be free from sharp edges and well polished.

4-39 Deforming tee nut with the vise.

4-40
Double terminal assembled.

As a rule the negative terminal is slightly larger than the positive terminal, since negative charges leak off more easily.

Charging the generator

In Fig. 4-41, Diana shows how to charge the spinning discs. First, note the white arrow on the disc, which indicates the motion of the front disc; the back disc moves in the opposite direction. You can see how to place the neutralizers: they form an X, and one blade from the backside is next to the head of the arrow. The angle between the two topside blades should be about the same as is shown in Figs. 4-42 and 4-43.

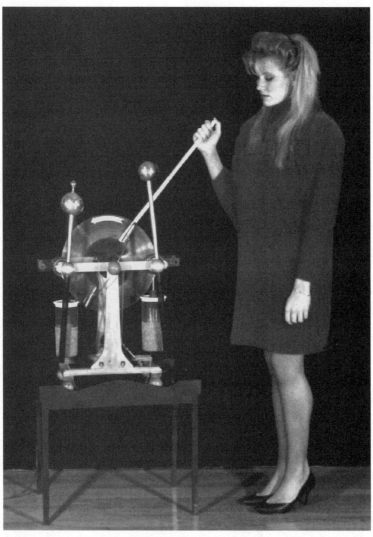

4-41 Charging the generator with PVC pipe.

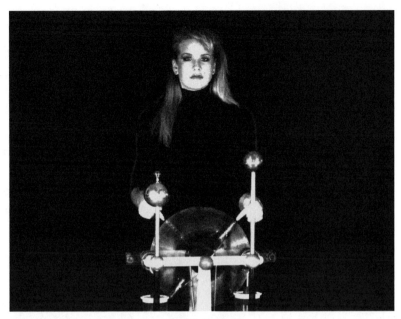

4-42 Neutralizers at widest angle: 90 degrees.

The charging tube is a 2-foot length of ½-inch PVC pipe. Rub the pipe briskly with a dry clean cloth until it crackles and place it as shown. In Fig. 4-44, seen from the end view, the charged tube is placed parallel to the neutralizer blade and behind both discs, about ¼ inch away.

4-43 Neutralizers at smallest angle: 40 degrees.

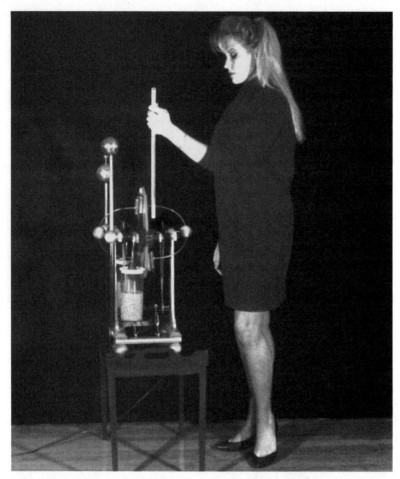

4-44 Charging; place PVC pipe on backside of both discs, next to neutralizer blade.

As shown in these photos, the charged PVC tube will impart a negative charge to the single 1¾-inch ball and a positive charge to the double-ball terminal. Later, when condensers are added, this polarity will produce the longest sparks and highest voltage.

Note: For safety's sake, work on dry and clean wood floors, and stand on a solid black rubber "welcome" mat 18 × 30 inches. Wear shoes with rubber soles, but without nails.

When the machine is charged, its speed will drop slightly and produce a crackling sound; the smell of ozone will fill the air (which you should remove with adequate ventilation). Bringing the terminals together, you should get a quiet brush-shaped discharge about 7 inches long in the darkened room. If the room is very humid (over 80 percent, for example) use a dehumidifier to lower the relative humidity to 50 percent for best results.

When you have experimented for several weeks without being shocked, proceed to build some discharge tongs and Leyden jars.

Discharge tongs and safety

The discharge tongs are essential for safely removing charges from Leyden jars used on the Wimshurst. Never use wires or screwdrivers. Figures 4-45 and 4-46 illustrate the homemade tongs.

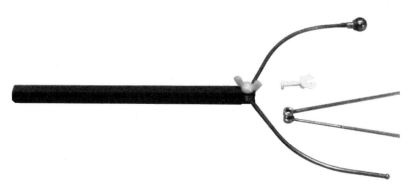

4-45 Homemade discharge tongs.

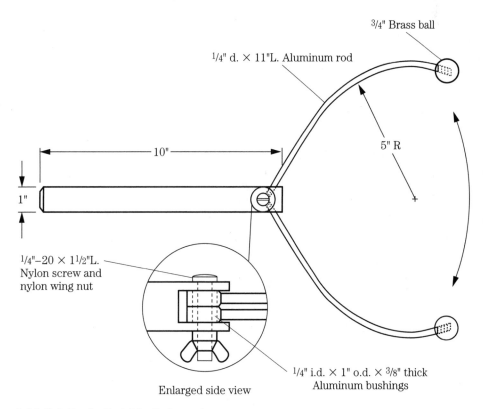

³/₄" Brass ball

¹/₄" d. × 11"L. Aluminum rod

10"

1"

5" R

¹/₄"–20 × 1¹/₂"L. Nylon screw and nylon wing nut

¹/₄" i.d. × 1" o.d. × ³/₈" thick Aluminum bushings

Enlarged side view

4-46 Details of adjustable discharge tongs.

The PVC handle, which is hexagonal or square in shape, should be solid, not tubular, to prevent dust collection inside. The curved $\frac{3}{16}$-inch brass wires are silver-soldered into $\frac{1}{4}$-inch i.d. shaft collars and they can be adjusted with the $\frac{1}{4}$-inch nylon screw and wing nut.

Leyden jar condenser set and hookup

The Leyden jars act like modern electrical capacitors, providing energy storage. The term *condenser* was originally used because at the time, electricity was pictured as being like a fluid substance, so the condenser was essentially a storage tank. Condensers are needed for sparks.

I made the jars in Fig. 4-47 using a pair of standard 1-pint Rubbermaid drinking tumblers with covered lids. In general, food storage containers made from polyethylene work better than other plastics because they are puncture-resistant. For this size Wimshurst generator, the Leyden jars should be about 3 inches diameter × 5½ inches tall. The lids are friction-fit.

4-47
Leyden jar condenser set with chain link.

For each jar, bore a $\frac{5}{8}$-inch-diameter hole in the center of the lid and turn the plastic bushing for a press-fit in the lid as shown (Fig. 4-48). Bend the jar hangers from $\frac{3}{16}$-inch brass rod or copper ground wire into the shape given (Fig. 4-49).

The overall length of wire is about 9¼ inches and the finished height of the wire hanger is 8 inches. I first make a pattern of the hanger with a soft wire of solder, then

4-48
Jar lid with plastic bushing, friction fit.

match this with the stiff brass. I drill a small depression in the top end of the hanger and tin with silver solder, invert this end, and place a ⅜-inch steel or brass ball with a bit of flux on the tinned end. Slowly heat the ball to a brown color (if it's steel) until it settles into the depression. Cool with water and test for strength. In the opposite end of each hanger, drill a ³⁄₃₂-inch hole. De-burr and polish these hangers—especially the

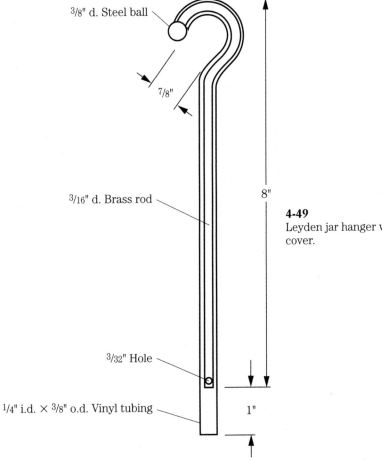

3/8" d. Steel ball

7/8"

3/16" d. Brass rod

8"

3/32" Hole

1/4" i.d. × 3/8" o.d. Vinyl tubing

1"

4-49
Leyden jar hanger with vinyl tube cover.

ball terminals—clean with alcohol, then slip on clear vinyl tubing ¼ inch i.d. × ⅜ inch o.d. × 10 inches long, so that the tubing extends about 1 inch below the rod and jewelry chain connection. The chain should extend another 3 inches. Slip on the bushing and lid for a friction-fit.

Mask the tumblers around the outside above the height of 2½ inches. Roughen the bottom 2½ inches with 400-grit sandpaper so glue will stick. Repeat this for the inside surface, then clean tumblers with alcohol. Using a paper pattern as shown, cut two outside coatings from 0.002-inch brass or steel shimstock (Fig. 4-50). This goes around the tumbler circumference with ⅜ inch extra for an overlap joint. The outside bottom of the jar can be coated with a disc of *extra heavy* aluminum foil cemented with superglue.

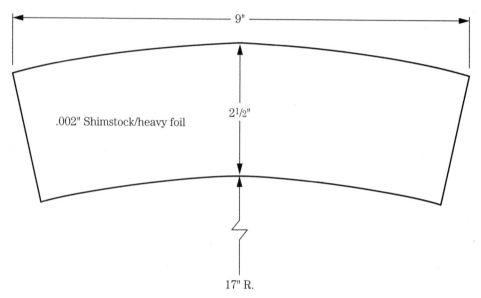

9"

.002" Shimstock/heavy foil

2½"

17" R.

4-50 Pattern for jar's shimstock coating.

When the bottoms are coated, wrap the shimstock coatings tightly around the outside and cover the vertical lap joint with a strip of sheet polyethylene and over this lay a splint of wood ½ inch square flush with the edge of the lap joint. Clamp the sandwich down onto the tumbler using a deep throat C-clamp (Fig. 4-51). Run superglue into joint; when set, peel off the polyethylene strip and remove excess glue with acetone. This completes the outside coating. Use same pattern to glue on extra heavy aluminum foil inside and use wood glue on the roughened surface for the cement. Smooth down wrinkles and bubbles, including the bottom. When installing the lid, be sure the hanger chain reaches to the foiled bottom.

The pair of Leyden jars are electrically connected together with a 24-inch length of jewelry link chain. The ends are taped to the jar-foiled bottom on the outside using a section of adhesive aluminum foil. When installed, the middle of the chain hangs down and touches the table to help ground the outer coatings.

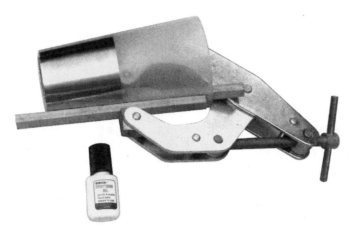

4-51 Clamping shimstock coating for gluing.

Note: The two Leyden jars working together can store up to a 9- to 10-inch spark or roughly 150,000 volts each. They should be treated with the same respect you would show for blasting caps and dynamite. When finished with the generator, discharge across the terminals and also discharge each jar to its outside coating and remove both jars when not using for safety's sake (Fig. 4-52).

4-52
Miniature lightning bolt.

A large Wimshurst for the university physics lecture room

What follows is a brief description of a sectorless Wimshurst with 24-inch discs; it is simply a scaled-up version of the smaller one just described (Fig. 4-53).

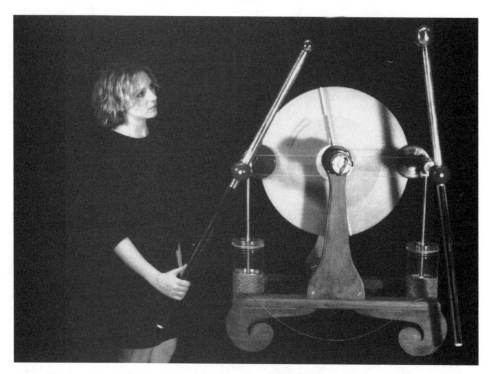

4-53 The lecture room Wimshurst.

Special note: I encourage only well-seasoned experimenters to build this powerful generator. Building the small Wimshurst first will make you familiar with the construction methods and hazards and materials involved. Review each step for building the 12-inch Wimshurst generator to fill in the details. In most cases, dimensions and spacings are doubled. Design of this generator centers around the direct-drive motors available.

The stand and upright supports

The two side rails are cut from 2 × 4 lumber and are 32 inches long. Overall width of the stand is 18 inches, and the inside distance between the upright supports is 11½ inches. (Refer back to Fig. 4-2.) The uprights (Fig. 4-54) are each composed of a sandwich of three 1 × 10s, this greater thickness being needed to support the extra weight of the motors and discs. I divide the middle 1 × 10 lengthwise to form the ½-inch slot for the wiring raceway and slide it down 1½ inches to form the split tenon below. Remove a section at the top of this board to make the 3-inch slot for the crossbar. The sandwich is glued together, cut out on the bandsaw, and the edges are rounded with the router. The top of the uprights must be large enough for mounting the motors; the 2-inch hole for accommodating the motor shaft is centered in the slot.

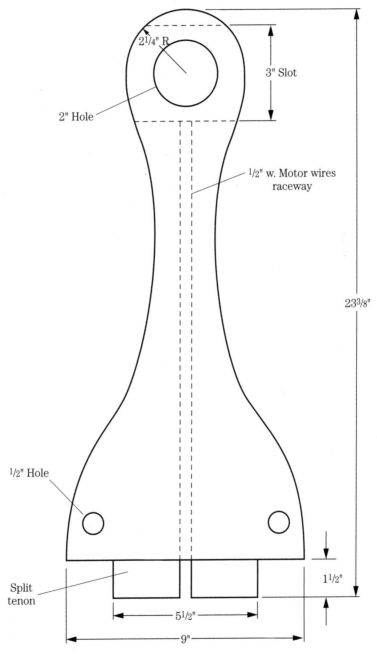

4-54   Upright support pattern for 24-inch-disc Wimshurst.

Motors and mounting

Because of the large diameter of the discs, I chose high-torque motors to overcome the electrostatic resistance created. The matched pair of motors consisted of Von Weise brand gear motors. These reversible motors run on 12 to 90 Vdc with variable speeds of up to 340 rpm and are continuous-duty with maximum torque of 13 inch-pounds. Dimensions are $3\frac{1}{2} \times 3\frac{1}{2} \times 7$ inches and each weighs about 7 pounds. The shaft size is $\frac{1}{2}$ inch in diameter $\times$ $1\frac{3}{4}$ inches long; because the shafts are too short, I turned shaft extenders from aluminum to give an extra $3\frac{1}{2}$ inches for better disc support. Solder extensions onto the motor wires to pass through the raceway in the upright. A small hole is drilled just below each motor in the upright to pass the wires inside (see Fig. 4-12). The square motor face plate is secured to the upright with four machine screws. Ground one of these screws (each upright) to protect the motors.

The discs

These must have good dielectric properties and low moisture absorption, be flat, and have uniform thickness. One choice is G-10 epoxy-glass; an even better would be *cast* delrin, which is expensive but has an even thickness, which means less vibration at higher speeds. Each disc measures $\frac{3}{16} \times 23\frac{3}{4}$ inches.

Bosses and bushings

Figure 4-55 shows the finished product; collar and bushing are turned from delrin. The bosses may be turned from $\frac{3}{4}$ inch gray or white PVC plate or polypropylene. Remember to take the final face cut by mounting bushing and boss on $\frac{1}{2}$- inch drill rod and turning with tailstock center. This prevents disc wobble. Each disc is secured with (eight) $\frac{1}{4}$-20 $\times$ $\frac{3}{4}$-inch flathead *nylon* screws countersunk into the disc face. The discs are ready for edge truing (see Fig. 4-14).

Neutralizers and mountings

Figure 4-57 shows an aluminum split collar that allows for easy removal of the neutralizer arms and blades and for locking neutralizers at a desired angle.

An aluminum ring $1\frac{3}{4}$ inches long $\times$ 2 inches o.d. is epoxy-glued into the upright (section view AA' in Fig. 4-58). Before the glue sets, be sure that the ring is perpendicular so that the neutralizer blades are equidistant from the disc face. The brass neutralizer rods or arms are $\frac{1}{4}$ inch in diameter and are bent as in Fig. 4-17 so as to spread $23\frac{3}{4}$ inches apart. The scaled-up neutralized blades are 4 inches long cut from aluminum or brass shimstock. I've found it helps to ground the neutralizer collar with a small jumper wire to one of the motor mounting bolts.

Power supply

Gear motors draw little current, about 1.1 amps for these 90-Vdc motors. You can use a variac or rheostat to step down the 110 Vac and then pass this through a full-wave bridge rectifier (Klaus Radio part # NTE 5309), for example. Be sure to attach the rectifier to an aluminum heat sink and ground the heat sink. This prevents high voltages from sparking to the rectifier and burning it out.

4-55 Disc boss and bushing, mounted.

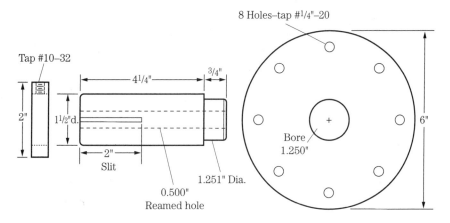

4-56 Locking collar, shaft bushing, and boss.

4-57 Neutralizer split collar.

Charge collector support system

The critical dimensions of the support bar cut from ¾-inch acrylic plate are shown in Fig. 4-59. The 1.65-inch holes are approximate and accommodate standard PVC 1¼-inch to 1-inch reducer bushings and 1¼-inch PVC caps.

Deepen the threads on the reducer bushing, and cut four slots in the threaded end to create a clamping action on the aluminum collector tube when the cap is tightened. (See Figs. 4-26, 4-27, and 4-60.) The bushing and cap are bored to give a snug fit for the 1¼-inch o.d. collector tube, about 1.251 inches. Each aluminum tube is 15⅞ inches long. The wooden ball sockets for holding the discharge handles (see Figs. 4-33, 4-34, and 4-35) are 3 inches diameter with a thru hole 1¼ inches diameter.

When finished machining the ball sockets, smooth the edges of the holes until they are slightly rounded, clean well, and give each three thin coats of red high-voltage varnish (available from Klaus Radio, Inc.). The coating must be smooth and without pores in order to reduce coronal leakage.

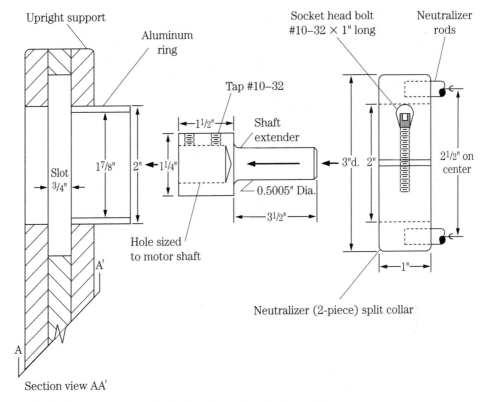

4-58 Upright, ring, motor shaft extender, and neutralizer collar.

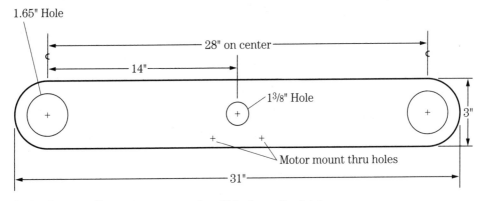

4-59 Charge collector support crossbar (¾-inch acrylic plate).

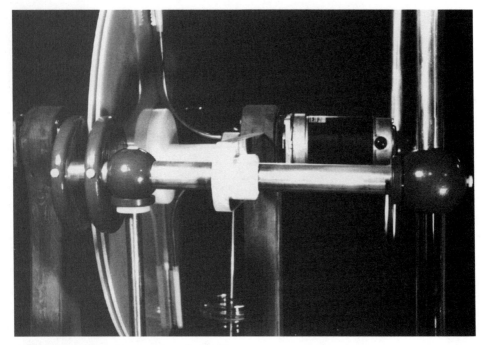

4-60 Aluminum collector tube, mounted.

Charge collector combs

Figure 4-61 presents an alternative method similar to Winter's collector comb rings (see Fig. 2-1). Mine are wood discs ¾ × 5½ inches in diameter with edges well-rounded. The inside surfaces have grooves routered out to a depth such that ½-inch thumb tacks may be cemented in a circular arc with points flush with the wood disc's surface. The grooves are lined with heavy aluminum foil with a single ¼-inch-wide strip of copper foil connecting the tacks to the aluminum tube. The wood combs are smoothed, carefully cleaned, and given three coats of red varnish. Each comb is locked in place with a single nylon screw. Note that the outer terminal comb has a *blind* hole 1¼ inches diameter so as to prevent serious leakage from the end of the collector tube.

Refer also to Figures 4-29 and 4-30 showing PVC caps modified for use as combs. The 24-inch Wimshurst will need 6-inch PVC caps; caps that are domed rather than flat-ended will have a more pleasing appearance. Remember that all parts of the collector system should be smooth, clean, and well polished to reduce high-voltage leakage.

Discharge handles and terminals

As shown in Figs. 4-31 and 4-32, the acrylic handles are 1¼ inches in diameter × 24 inches long, and the aluminum tubes are 1 inch i.d. × 1¼ inch o.d. with the longer tube being 21 inches and the shorter tube 18½ inches long.

4-61   Wooden collector combs, opened.

The positive double terminal, as shown in Fig. 4-38, should be approximately 1¼ inches in diameter for the top, and the larger hollow ball should be 3 inches in diameter. Chinese exercise balls and float balls with butt-welded seams make good terminals. The larger negative terminal may be a 4-inch-diameter float ball that simply slips down over the aluminum tube.

Leyden jars

The Leyden jars may be hung with hooks as in Fig. 4-47 or joined with metal tubes running straight up into wood sockets at the collector tube (see Figs. 4-60 and 4-62). The jars are acrylic food storage containers 5¼ inches o.d. × 9½ inches tall. The connector tube passing up to the collector is ½-inch copper inside with ¾-inch butyrate plastic tubing slipped over to reduce coronal leakage.

Properly constructed, clean, and well polished, the 24-inch Wimshurst should give sparks up to 16 inches long, as seen in Fig. 4-63.

4-62 Leyden jars from food storage containers.

Troubleshooting and general maintenance

To see the luminous effects produced by the generator, completely darken the room and after the eyes have a chance to adjust (about five minutes), notice the area around the discs. The right side of the polarity diagram shows the luminous appearance of the charged generator (Fig. 4-64).

Observing the generator in the dark not only produces a beautiful light show, it can also help pinpoint leaks (they show up as tiny spots of light) in the collector system and the condenser surfaces.

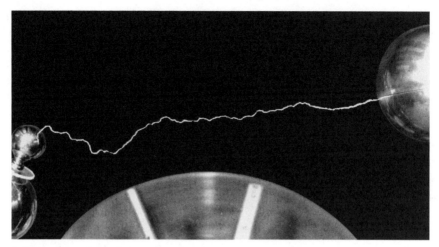

4-63   A 16-inch spark from 24-inch discs.

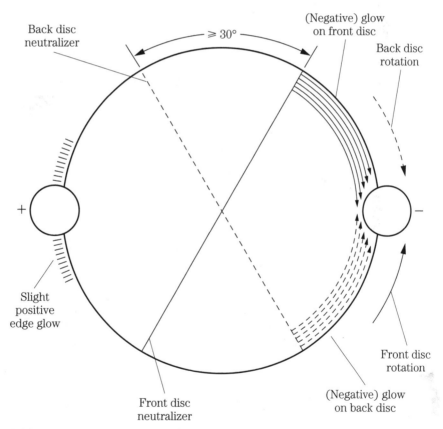

4-64   Generator polarity diagram.

Clean the generator parts as described in Table 4-1, using alcohol and a lint-free cotton cloth. Occasionally clean the Leyden jars and check for pinpoints of light in the dark to locate any charge leaks. Because there are no rubbing parts in a sectorless Wimshurst, there is little to wear out. Motors and bearings should last indefinitely. Check disc spacing and neutralizer blade air gaps for proper setting and try to operate generator at normal room temperature with relative humidity at less than 70 percent.

Note: Do not perform electrical experiments with the very young, the elderly, infirm, or those needing pacemakers. Do not use or store sensitive electronic equipment in the same area. Always discharge the generator with discharge tongs and store Leyden jars separately.

Table 4-1. Troubleshooting chart

Problem	Solutions
1. The generator is difficult to charge	1a. If one neutralizer blade doesn't take a charge, try the other blade. (For some reason, this often works!)
	1b. Check the disc and neutralizer spacing. The discs should be very close together (but not touching), with a gap typically about ⅟₁₆". The space between the neutralizer blades and the discs must also be about this distance.
	1c. The discs might be dirty. Clean the disc surfaces using a clean, lint-free, absorbent cloth moistened with denatured alcohol. Without removing the discs, turn them by hand and clean the outer disc surfaces by working from the center to the edge. It is not enough to just blow off the visible dirt or dust, since moisture and chemical residue may be invisible. Let dry thoroughly.
	1d. The room temperature might be too cold or the air too humid.
2. The charge dies out after a large discharge or when the terminals touch.	2a. The angle between the neutralizer blades might be too narrow. Adjust them so they are more than 30° apart.
	2b. The air may be too humid.
3. The generator ceases to charge after being turned off, but the discs are still turning (slowly).	3a. The air may be too humid.
	3b. The discs might need cleaning.
4. Sparks jump from the collectors to the neutralizer blade.	4. The neutralizer blades might be too far apart and too close to the collectors.
5. In a darkened room, silent and luminous points of light appear in the collector comb area.	5. Clean the collector system (including the terminals and handles) to remove dust, lint, and hair.
6. In a darkened room, points of light come from the terminals or handles.	6. Check all the metal surfaces for nicks and scratches (by sight and touch). Only use a crocus cloth (a type of sandpaper) for polishing, never steel wool.

Table 4-1. Troubleshooting chart (*Continued*)

Problem	Solutions
7. Sparks appear on the sides of the Leyden jar.	7. Carefully discharge the jar using tongs. Clean inside and outside of the jar and its lid and hanger.
8. The generator, with jars attached, charges, but gives only feeble sparks.	8a. Check the grounding chain to make sure it joins both jars.
	8b. A jar might be ruptured. In the dark, this is evidenced by a short, continuous bright spark in the coated portion of the condenser.
	8c. The angle of the neutralizer blades might have inadvertently been reversed. Refer to the polarity diagram to check

5
CHAPTER

Unusual generator designs

Earlier I mentioned an advanced influence machine that was much more efficient than the Wimshurst machine: the Wommelsdorf generator (covered by U.S. patents 882,508 in 1908 and 1,071,196 in 1913). So, if this machine was such an improvement, why is it not as well known as the other designs? Mainly because of poor timing. From 1900 onward, the momentum in electrical engineering shifted toward electromagnetic technology because of its practical power output. By the time the Wommelsdorf machine made its debut in *Electrical Review* in 1913, there were few articles on the subject. Also, the plastics "revolution" had not yet occurred; ebonite and Bakelite were the best insulators on the market, neither of which are resistant to ozone attack.

This machine is an example of how technical innovations fall into oblivion because of bad timing, unfavorable economic conditions, or the lack of materials to make the device practical. This is why it is always a good idea to review older innovations and reassess them in light of our present situation. There are quite a few "jewels in the rough" in any branch of engineering that are still waiting to be rediscovered.

The article in Fig. 5-1 is intended to supplement the two Wommelsdorf patents. Table 5-1 gives generator performance for several sizes available about the year 1915.

The popularity of this design continued in Germany into the 1930s. Figure 5-2 shows a commercially produced portable unit, probably used by electrotherapists.

Improved designs

Since the Wommelsdorf machine was invented, three French innovations have improved disc generators. The first of these innovations by Henri Chaumat; his single-disc generator (46 cm diameter) produced a potential of 300,000 volts. The spark length was 32 cm between terminals which were 26 cm in diameter. He claimed its output was better than a Holtz or Wimshurst machine's—24 watts at 200,000 volts. The Wimshurst generator delivered 0.7 watts at 70,000 volts. No U.S. patent was received by Chaumat.

*The condenser machine

The condenser machine invented by Dr. H. Wommelsdorf is a new electro-static machine for the direct generation of high-tension continuous current. While its principle was described some years ago in the scientific press, it has only recently been so improved as to become suitable for practical use in X-ray and other work.

The characteristic feature of this machine is the alternating arrangement of the rotary and stationary disc, which is based on the condenser principle. It will be remembered that in an influence machine the rotating disc is only influenced on one side, the generated electricity being drawn off from the other. The rotating discs of the condenser machine, on the other hand, undergo electrostatic induction on both sides, the electricity produced in them being collected from a groove in their extreme periphery by steel wires penetrating therein. It will thus be seen that each disc, in accordance with the theory of condensers, takes up and supplies twice as great an amount of electricity as an influence machine. Actual experience moreover, shows that the output of the new machine, thanks to some additional advantages connected with the condenser principle, is even considerably higher than could be expected. It is mainly due to the close arrangement of the discs that the condenser machine yields an amount of electricity 20 to 50 times as much as a Holtz-Wimshurst multiple-disc influence machine of equal size, the type so far almost exclusively on the market. Another point of considerable importance is that the condenser machine is absolutely independent of atmospheric conditions.

The sectors of the condenser machine are not attached to the surface of the discs, but embedded in their interior, thus augmenting the output, pressure and self-excitation of the machine, and increasing the life of the discs. In fact, the unceasing inflow and outflow of high-tension electricity from the suction combs, brushes and sectors to the insulating discs was bound in a very short time to destroy even the best insulating material. The stationary discs of the machine are likewise entirely encompassed by insulating material.

The condenser machine mainly comprises a substantial frame closed on all sides which carries in its interior the statical fields, that is, the stationary discs. The rotating discs above described, to which is due the influence effect, are sandwiched in between these fields, thus being also comprised in the interior of the frame. This compact arrangement protects all vital parts of the machine not only against dust, but at the same time against radiation losses which would otherwise be an important item.

That this remarkable simple and efficient machine—as it were, a continually charging and discharging condenser—should have had to wait a number of years before being placed on the market, was due to some drawbacks of its original construction and to the fact that no insulating material suitable for the purpose could be found. In fact, the effects of the electricity and ozone produced by the condenser machine are, just because of its extraordinary output, so powerful that rubber, a substance that could not be dispensed with on account of the high voltage of the machine, would become conductive after a very short time.

The adoption of a certain brand of Bakelite with which the caoutchouc discs were coated all round, first made the condenser machine durable and suitable for commercial purposes. This Bakelite layer—insoluble and brilliant as enamel—endows the discs with a remarkably hard surface, and can only be removed by scraping with a knife, in the form of an amber-like yellow powder.

The arrangement of the current collectors at the extreme circumference of the discs has enabled the spark length of the condenser machine to be

* From *The Electrical Review*, August 29th, 1913.

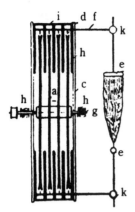

a, Rotating discs; c, end cover; d, collector; i, frame;
e, electrodes.

SECTION OF CONDENSER MACHINE.

increased to twice its former value, thus allowing sparks considerably longer than half or even two-thirds of the disc diameter to be produced.

The condenser machine will be found an extremely useful apparatus for X-ray work, affording, as it does, a means of producing high pressure continuous current directly without any conversion, rectification, &c. This allows hand operation to be used in the case of moderate effects, while increasing the life and constancy of X-ray bulbs, the more simple of which can be used without cooling, thus cheapening operation. Wherever there is no suitable electricity supply, *e.g.*, in the case of portable installations for military and other purposes, the condenser machine will afford an ideal means of operating X-ray apparatus. It will also be found useful in cases where only an alternating or three-phase current supply is available, which could not be used without resorting to converters, rectifiers, &c. Apart from hand operation, a small continuous, alternating, or three-phase current motor (of $1/5$ H.P.) can be used to drive it. The output of a motordriven condenser machine having a single rotating disc 55 cm. in diameter is as much as $1/2$ milliampere. Machines comprising four, eight or ten rotating discs, such as have been constructed, show a correspondingly higher output.

Another field in which the condenser machine

5-1 Continued.

is found to prove of the greatest usefulness and to excel all other apparatus so far in use, is the field of electro-therapeutics, viz., the medical applications of electricity. An especially important feature in this connection is that the output of the machine can be controlled at will and reduced to zero, the current intensity being proportional to the number of turns per second. This is an enormous advantage over powerful induction coils, which cannot be used for small outputs.

If, for instance, the electric currents yielded by the machine be applied to the human body, their physiological effects can be altered quite gradually and controlled at will, from a hardly perceptible shiver up to an unbearable intensity. At the same time, the operation of the machine is absolutely safe and harmless, the upper intensity limit being readily adjusted for by choosing a proper size of Leyden jars.

The machine also comprises two special terminals from which alternating currents or rapid oscillations can be collected for high-frequency current work and for performing the methods devised by Oudin, D'Arsonval, Apostoli, &c. The same large condenser machine also lends itself for franklinisation proper, that is, for static electricity treatment, and for experiments with the insulating stool.

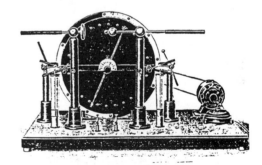

THE CONDENSER MACHINE.

The high pressure of the machine is noticed at a considerable distance by a spider-web feeling due to electrically-charged particles. These remote effects can be strikingly demonstrated by lighting Geissler and Tesla tubes from a great distance.

The exceptionally high Leyden jars used in connection with the condenser machine are countersunk in its base, and are, like the discs, coated with Bakelite. The polariser at the same time serves as short-circuiter, for rapidly and accurately breaking the flow of sparks.

Though the machines so far constructed are doubtless the starting point of types much larger and more powerful, they are able to comply with the most exacting requirements of X-ray work and electro-therapeutics.

The condenser machine is constructed by Messrs. Berliner Elektros. G.m.b.H., of Schonberg, near Berlin.

5-1 Continued

Table 5-1. Wommelsdorf generator performance*

No. of rotating discs	Diam. in cm	Length of sparks in mm	Current in microamps	Energy required in H.P.
1	26	170–190	120–140	about 1/16
1	35	210–250	230–280	about 1/8
1	45	260–300	380–480	about 1/6
1	55	300–360	500–600	about 1/6
2	45	260–300	620–750	about 1/6
2	55	300–360	700–850	about 1/4
3	55	300–360	1,400–1,600	about 1/3
5	55	300–360	2,000–2,500	about 1/2
7	55	300–360	2,800–3,400	about 3/4
10	55	300–360	3,750–4,500	about 1

*(Generator speed for the above was 2,300 rpm.)

5-2 Improved Wommelsdorf. *Elektrotechnische Zeitschrift,* 1929

The second generator design was developed from Chaumat's prior work and was the invention of Pierre Jolivet in 1953. Jolivet's disc generators used bleeder resistors, by which a portion of the output was used to increase the electric field intensity across the inductors—called *neutralizers* in a Wimshurst machine. No U.S. patent was found for Jolivet's invention, and neither of the previously mentioned articles has been found translated from the original French into English.

The last French innovation appeared concurrently with Jolivet's in the early 1950s. The inventor, Noel Felici, chose to concentrate wholly on cylindrical induction generators and managed to raise the generator efficiency to a remarkable 90 percent, using an environment of pure hydrogen at 15 to 20 atmospheres. A drum generator with a power output ranging from 20 to 3,000 watts was feasible with Felici's design. His discoveries so dominated the study of electrostatics that most of the generator patents issued in the United States during the 1950s went to this man. Felici's design is somewhat exotic, and therefore is outside the scope of this book, but it does show the room available for improvement in performance.

Moving in another direction, I close this chapter with several novel and unique high-voltage generators.

Sparks from condensing steam

In the autumn of 1840 at Seghill, England, a mechanic experienced small shocks when he stood in the jet of steam leaking from an engine boiler and touched the safety valve. Science experimenter Hugh Pattinson investigated further and was able to get ⅜-inch sparks and also charge a Leyden jar. A rival investigator, William George Armstrong of Newcastle, also took up the challenge and experimented with boilers. He concluded that the electricity was not produced in the boiler or its tubing but that the steam became electrified when it entered the atmosphere.

His specially constructed boilers, insulated on glass legs, were called "hydroelectric machines" (Fig. 5-3).

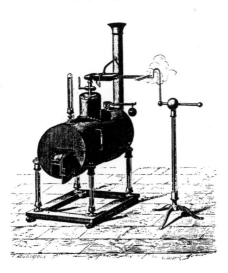

5-3
Armstrong's hydroelectric machine. *Ganot's Physics, 1906*

A popular textbook[1] description follows:

The Armstrong generator produced charge by the frictional impact of condensing steam in the brass and wood nozzle. The hole size was ⅛ inch and the boiler was operated at 60 pounds per square inch pressure.

With 46 nozzles, the noise was quite loud but current and voltage were much greater than existing (frictional) high voltage generators. It could charge a Leyden battery of 33 square feet to capacity 60 times per minute where a large glass disc generator required a minute per charge.

One small hydroelectric generator still exists at the Newcastle Museum of Science and Engineering.

The persistent attention to a small curious phenomenon led Mr. Armstrong into a largely uncharted area of electricity; this should inspire today's electrical experimenter.

Kelvin's waterdrop electrostatic generator

In June of 1867, William Thomson (Lord Kelvin) described a high-voltage generator that develops a potential difference from falling drops of water (Fig. 5-4). This is an induction-type generator and the potential is normally limited to 15,000 volts. This could be improved on using pressured jets and better insulation. The operating principle is given here[2]:

A and B of Fig. 5-4 are insulated metal cylinders called the *inductors*, and are in metallic connection with two cylinders a and b, also insulated, called the *receivers*, each having a funnel, the nozzle of which is in the centre of

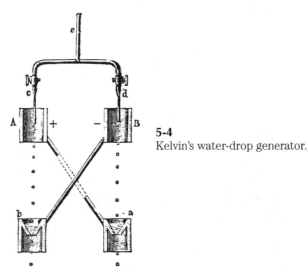

5-4
Kelvin's water-drop generator.

---

1 Source: *Elementary Treatise on Physics* by Adolphe Ganot (1906), translation by E. Atkinson, Wm. Wood & Co., N.Y.

2 Source: *Elementary Treatise on Physics* by Ganot.

the cylinder. Water from the pipe *e* falls in drops through the metal taps *c* and *d*, the nozzles of which are in the centre of the cylinders A and B.

Take first the case of cylinder A, and suppose it to possess a small negative charge at the outset. The drops as they fall will be charged negatively by induction, the corresponding positive going to earth, through *e*. Falling on the funnel of the receiver *b*, they impart to it the whole of their charge, and the water as it issues will be neutral.

But the negative charge of B is shared with *b*, which is thus negatively electrified, and the drops which fall though it are positively electrified and give up their positive charge to *a*, which strengthens the positive of A. By this means even with a very slight original charge they will strengthen each other, until even sparks pass. It is not even necessary to give a charge at the outset; the ordinary electricity of the atmosphere is sufficient.

The energy in this apparatus is derived from that of the falling body, and would be exactly equivalent to it if there were no loss, and if the drops reached the funnel without any velocity.

General plan for a Kelvin waterdrop generator

The basic parts and dimensions are given in Fig. 5-5. The framework of wood supporting these parts is not shown. A rod clamping onto rubber tubing regulates both jets. This horizontal rod has an eye at each end with a wing nut and screw for controlling the pressure against the vertical tubes. The two nozzles are plastic or glass tubes with a tip hole of about ¹⁄₁₆ inch. The two rings below the tips are cut from copper or brass tubing 1 inch in diameter, with edges smooth or round steel rings used with rope and tie-down lines at the hardware store. The rings are soldered to rods stuck into holes in an acrylic strip behind.

The aluminum plates below the two cans are well insulated on an acrylic base plate. The aluminum plates and cans are cross-connected with well-insulated sparkplug wire with the metal rings and are hooked to the rods behind the rings next to the plastic strip (so that they do not distort the water jets). To reduce charge leakage due to water spray, keep the aluminum plates elevated ¼ inch above the acrylic plate.

Note that the two jets are fed from one tank above and are gravity fed. The drop rate has been measured at about 150 drops per second.

Figure 5-6 shows a device described in 1912. When I made a modified form of this machine, I substituted two 4-inch-diameter float balls for the discs to reduce leakage. I joined them together to make a dumbbell shape, and used nylon thread in place of linen. A small dc motor and pulley drive the dumbbell at 6000 to 8000 rpm.

I used my simple generator to study the influence of static electricity on vortices in air. At full speed, I lower the dumbbell terminal into a wastepaper basket partly filled with Styrofoam packaging "peanuts." Charging does increase the strength of the miniature tornado vortex that forms. My unit also produces sparks ¾ inch long.

This generator is really like a Van de Graaff generator in its simplest form. An interesting science fair project would be to compare different pulley metals with various kinds of threads—the speed and tension remaining fixed, of course. There should be a variation in potential for each combination of metal and type of thread.

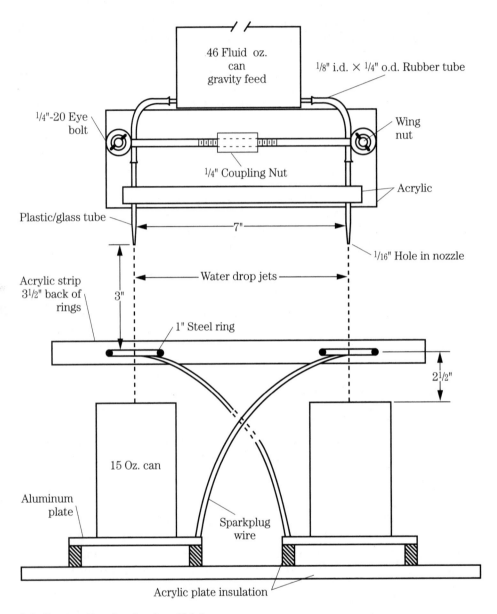

5-5 Construction drawing for a Kelvin.

The last unusual generator is one that develops high voltages through the impact of dust particles in a closed system (see Fig. 5-7).

Several impact generators have been patented over the years; some use steam or semiconducting liquids. (See U.S. patent 2, 078, 760 by Clarence W. Hansell, April 27, 1937.) Voltage and current will vary with air pressure, collector design, particle material, and size. Experimenters wanting to research along this line should provide a way to keep the compressed air dry.

Odd static generator

The outfit here illustrated I accidentally discovered will generate static electricity.

Upon a shelf or the edge of a table fasten a small motor that will run at a speed of about 3,000 revolutions per minute. On the shaft of the motor there should be a small grooved brass pulley. Provide a second pulley about ³/₄ inch in diameter and a small stove bolt that just fits the hole in it. From thin sheet copper cut two disks exactly alike, and six inches in diameter. Drill a hole in each disk of the same size as the hole in the pulley, and fasten the disks, one on each side of the pulley with the stove bolt as shown. Now make a belt of heavy linen thread. Place the thread over the pulley on the motor and set the disks' pulley upon the loop, thus leaving the disks suspended in the air. Start the motor, give the disks a turn and as soon as they come up to speed, they will act as a gyroscope and keep balanced. Take a piece of copper conductor in the hand and bring the metal near the disks. A ³/₄-inch spark can be taken off. An incandescent lamp held near the disks will glow with a weird blue light.—E.H. SAMEN.

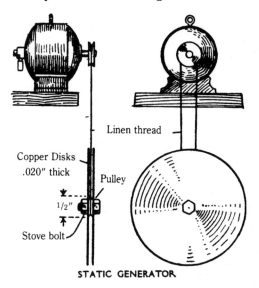

Linen thread

Copper Disks
.020″ thick

Pulley

1/2″

Stove bolt

STATIC GENERATOR

5-6 Simple generator. *Popular Electricity*, 1912

In Fig. 5-7, diatomaceous earth, also used in swimming pool filters, is hazardous when the fine dust is inhaled. Generators that rely on impact electrification come closest to duplicating nature's giant lightning displays. This method of charging is also responsible for fuel transfer and grain elevator explosions; frictional sparks in the presence of finely divided particles in suspension are the main hazard.

Experimental design modifications

Neutralizer units, also named *diametral conductors* or *inductors*, depending on theory of operation, are very important. Chaumat, Jolivet, and Felici explored this area extensively. Material, shape, disc area covered, and placement on the disc strongly influence generator efficiency. Some inventors inserted variable resistances between the arms of each neutralizer unit to see how conductivity affected performance. Jolivet, especially, used bleeder resistors of 200,000 to 700,000 ohms to skim off some energy from the collectors to intensify the inductor field strength across the disc.

Experimental disc materials

Some inventors received patents for their work on disc materials. For example, U.S. patent 937,691 by Burton Baker (1909) describes composite discs made from layers

A High Voltage Direct Current Generator

By Richard E. Vollrath
University of Southern California, Los Angeles

(Received August 19, 1932)

When powdered materials are blown through metal tubes by means of compressed air considerable quantities of electricity are produced by contact electrification. It was found that 6×10^{-5} coulombs could be produced per gram of diatomaceous earth, a form of silica, blown through a short length of copper tube. A generator of extremely high voltage is proposed, and a small scale model of such a generator is described, by means of which currents of 8×10^{-5} amperes at 260 kilovolts were generated.

THIS work was undertaken to provide numerical data to serve as the basis for the design of a high voltage generator capable of generating a milli-ampere at voltages above a million.

The Proposed Generator

The discussion to follow will be simplified by a consideration of Fig. 1 which is a diagrammatic representation of the proposed high voltage generator. The small scale model constructed will be described later on.

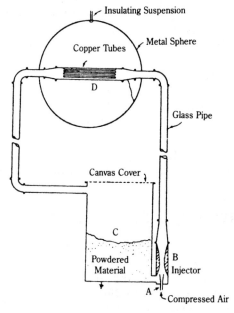

Fig. 1.—Proposed high voltage generator.

A blast of compressed air is blown from a nozzle *A* into a suitably designed injector *B*. A powdered material *C* is sucked into the air stream by the action of the injector and carried along through glass or Bakelite pipe. The air laden with the powdered material passes through a number of metal tubes *D* arranged in parallel within a large spherical conductor and electrically connected with it. The particles of powder become electrified by contact with the walls of the metal tubes. The charged particles are carried away from the sphere and returned to an earthed reservoir C. The potential of the sphere will rise until limited by corona discharge from its surface.

In order that one milliampere can be drawn from the sphere, enough powder must be blown per second to produce a charge of 10^{-3} coulombs or 3×10^6 e.s.u. per second. It was the main purpose of this work to find out if such charges can be obtained from reasonably small quantities of powder. From an engineering standpoint quantities up to about 305 grams per second should be feasible since sand blast machines have been constructed capable of blowing this quantity of sand.

Previous Work

It has been known for a long time that considerable charges are developed when particles of solids are blown over the surfaces of metals and other substances. Most of the work recorded in the literature gives no information as to the quantity of electricity produced in this manner by a given amount of material. However, the following brief resumé of the more pertinent articles showed that the charges obtainable were large enough to warrant further work along these lines.

According to Rudge[1] a few centigrams of flour blown into a large room produced a charged dust cloud whose potential as measured by a radium coated collector, or probe, was 200 volts. He pointed out that this and other results of a similar nature obtained by him accounted for potential gradients of 10,000 volts per meter observed during dust storms, and for lightening flashes occurring during the eruption of ashes from volcanoes.

Petri observed that a steel telegraph wire 5 kilometers long became electrically charged during a violent snowstorm. A continuous stream of sparks several millimeters long could be drawn from the wire; and Petri estimated the electrical power generated to be 1.2 horsepower. The effect has been attributed by Ebert and Hoffmann[2] to contact electrification of the snow blown over the wire by the wind.

A similar observation is recorded by Stäger[3] who exposed a wire 9 meters in length to the driving snow during a snowstorm. There was a distinct corona discharge around the wire and a current of 17 to 20 milliamperes could be drawn from it. In this case the power generated was estimated to be 3 watts. Stäger in the same article gave the charge carried away by hoarfrost blown from a surface of ice. Under particularly favorable circumstances it amounted to 1000 e.s.u. per gram of hoarfrost. He also mentions the appearance of a corona discharge 10 cm long during the production of carbon dioxide snow by rapid evaporation of liquid carbon dioxide escaping from a tank.

[1] W. A. Douglas Rudge, Proc. Roy. Soc. London A90, 256 (1914).
[2] Ebert and Hoffman, Meteor. Zeits. 317 (1900).
[3] A. Stäger, Ann. d. Physik 77, 230 (1925).

5-7 Continued.

It is quite likely that the tremendous voltages produced in the Alps, and lately used in attempts to operate large x-ray tubes are generated by the electrification of snow blown over the ice covered peaks.

Theoretical Limitations

In pursuing this work the writer adopted the views of Helmholtz on frictional electricity. According to these, so-called frictional electricity is developed whenever two dissimilar surfaces are brought into contact and then separated. A double layer of charges, whose magnitude is determined by the contact difference of potential between the two surfaces, forms at the surface of contact—one charge residing on one surface and an opposite charge on the other. When the two surfaces are separated the charges of the double layer are torn apart, and a charge remains attached to one surface while the other carries with it a like charge but of opposite sign. A contact of very short duration of two insulators followed by their separation suffices to produce considerable charges, which indicates that the double layer does not penetrate very far into the body of the insulators in contact. The thickness of such double layers has been estimated to be of the order of 10^{-8} to 10^{-7} cm. From this it is evident that in order to produce large charges by contact electrification large surfaces of contact are the main consideration. This immediately suggests that at least one of the two substances brought into contact should be in a finely divided state so as to present a large surface. In this case the charges are produced by blowing the finely divided material over a metal surface; for example the powder is blown through a metal tube. When the particles strike the metal surface and leave it they acquire a charge which they carry with them as they move along with the air stream. An opposite charge remains on the metal which, if insulated, rises in potential as long as the powdered material is blown over it.

Leaving out of consideration corona discharges from the conductor, the ultimate potential which can be reached depends upon the mobility, k, of the charged particles leaving the conductor and the potential gradient, X, at the point where they leave. The charged particles to escape must be impelled by the air stream with a velocity greater than kX. The electrical image force between the particle and the conductor is considered negligible owing to the smallness of the particles under consideration. It can easily be shown that the above requirement imposes no serious limitation upon the potentials attainable, even though the particles should have to overcome the maximum gradient possible in air, about 30,000 volts per cm.

A charged particle of radius r in air will have a maximum mobility when it is carrying the maximum charge q permitted by the limiting gradient at its surface, that is, $q/r^2 = 30,000$ or $q = 100\ r^2$ e.s.u. For particles of radius 10^{-4}, $q = 10^{-6}$. Consider 1 cc of material broken into approximately spherical particles 10^{-4} in radius, each charged with the maximum 10^{-6} e.s.u. The total charge carried by all the particles is

$$Q = 10^{-6}/[(4/3)\pi r^3] = -10^6/4.$$

5-7 Continued.

According to this only 12 cc of material would be necessary per second to carry a milliampere. The same calculation for particles of radius 10^{-5} cm gives $10^{-7}/4$ e.s.u. per cc. However, owing to the fact that air in very thin films has a higher breakdown strength, the gradient at the surface can be higher allowing it to carry a larger charge.

It now remains to show that a particle charged to the above calculated maximum can be driven against a gradient of 30,000 volts per cm. This can be done by making use of some data obtained by Deutsch[4] on the motion of charged particles in electric fields in connection with a study of the Cottrell process of precipitating dust from gases.

He found that particles of radius $r = 10^{-4}$, after having picked up a charge of 376 electrons $= 1.8 \times 10^{-7}$ e.s.u. in a corona discharge moved with a velocity of 0.56 cm/sec. in a field of 300 volts/cm. Particles of $r = 10^{-5}$ cm picked up a charge of 5×10^{-9} and moved with a velocity of 0.42 cm/sec. in the same field. If we assume that the mobility varies linearly with the charge on a particle, we can use these results to determine the velocity of a particle of $r = 10^{-4}$ cm and carrying the maximum charge (10^{-6} e.s.u.) in a field of 300 volts/cm. The velocity will be $v = (0.56 \times 10^{-6})/(1.8 \times 10^{-7}) = 3$ cm per sec. For the case of particles of radius 10^{-5} the velocity is less than 0.42 because the calculated maximum charge turns out to be less than that observed by Deutsch. With the above velocity of 3 cm/sec. in a field of 300 volts/cm, a particle of $r = 10^{-4}$, carrying a charge of 10^{-6} placed in a field of 30,000 volts/cm would move with a velocity of 300 cm/sec. Now a particle of this size can easily be blown with a velocity ten times as great. Apparently there is no difficulty to be expected in blowing the charged particles away from a highly charged conductor.

EXPERIMENTAL

The following experimental method was used to find the powder most suitable for the purpose in view. Fig. 2 shows the experimental arrangement.

Fig. 2.—Experimental arrangement for investigating powers for charges.

An insulated brass plate P was connected to one pair of quadrants of a Dolezalek electrometer and to a condenser c, as shown. One milligram of the powdered material to be investigated was placed on the brass plate, which was earthed and insulated before blowing off the powder with a puff of air. The magnitude of the charge produced was determined from the deflection of the electrometer and the capacity of the system. The powders were prepared by grinding various solids and sifting them through a a 300-mesh sieve. This could not be done very well with metals which were used as obtained in the form of considerably coarser powders. The materials studied were mercuric

[4] Deutsch, Ann. d. Physik **4**, 824 (1930).

5-7 Continued.

sulfide, mercuric iodide, sulfur, rosin, iron powder, antimony powder, clay, and diatomaceous earth, which is a form of silica occurring naturally in a very finely divided form.

The most promising materials were the metal powders and the diatomaceous earth. The metal powders could not be further investigated by the next method to be described because, owing to their great density, they could not readily be blown by the compressed air available. The diatomaceous earth turned out to be ideal for the purpose, not only because it gave such large charges, but also because it is very light and easily blown. It consists of particles 10^{-4} cm in diameter and smaller, and it can be obtained commercially at 50 dollars a ton.

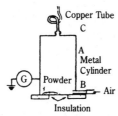

Fig. 3.—Experimental arrangement for measuring charge obtainable from diatomaceous earth.

The diatomaceous earth was used in larger quantities in such a manner as to permit the charges produced to be measured on a galvanometer. It was placed in a metal cylinder A, Fig. 3, 12 cm diameter and 30 cm high from which it was blown by means of compressed air introduced tangentially at B. The air laden with the powder passed through a piece of copper tubing C having an inside diameter of 0.5 cm and a length of 20 cm. The cylinder, insulated by standing on blocks of paraffin, was connected to ground through a calibrated galvanometer which indicated the current flowing from the cylinder as the powder was being blown out. The measurements were made by placing 5 grams of the powder in the cylinder and blowing it out with air flowing at the rate of one liter per second. The current is read on the galvanometer until all the powder is gone. The current is a maximum at the beginning of a run and decreases gradually as the amount of powder blown out per second decreases during the progress of the run. Since the current fluctuated somewhat, an average current was estimated during each half minute interval. These averages are listed in the second column of Table I which gives the result of a typical run. It should be noted that the average current recorded for the first interval is really too low because the initial swinging of the galvanometer prevents the current from being read at all during the first 15 seconds, during which time the current is considerably higher. A small current could still be read after the air had passed for 15 minutes. This is due to the fact that a small amount of the powder clung to the inner surface of the cylinder from which it was gradually dislodged and blown out by the air. The charge obtained per gram of powder is for these reasons somewhat higher than that given at the end of the table. The total charge in coulombs obtained from 5

TABLE I. *Charge obtained from 5 grams of diatomaceous earth blown by air flowing at rate of 1 liter/sec.*

Time in min.	Current in amp. $\times 10^8$	Charge in coulombs $\times 10^8$	Time in min.	Current in amp. $\times 10^8$	Charge in coulombs $\times 10^8$
0.5	137.5	4125	8.0	16.5	495
1.0	96.3	2889	8.5	13.8	414
1.5	68.8	2064	9.0	10.5	315
2.0	55.0	1650	9.5	8.3	249
2.5	41.3	1239	10.0	5.5	165
3.0	41.3	1239	10.5	4.7	141
3.5	41.3	1239	11.0	3.9	117
4.0	57.8	1734	11.5	3.0	90
4.5	82.5	2475	12.0	2.8	84
5.0	82.5	2475	12.5	2.2	66
5.5	68.8	2064	13.0	1.9	57
6.0	63.3	1899	13.5	1.9	57
6.5	46.8	1404	14.0	1.4	42
7.0	35.8	1074	14.5	0.8	24
7.5	27.5	825			

Total $30,696 \times 10^{-8}$ coulombs
or $\qquad 6.1 \times 10^{-5}$ coulombs/gram

grams of powder was found by adding together the product of current in amperes and time in seconds for all the half minute intervals. The charge on the powder is negative.

The copper tube C in Fig. 3 was at first straight, and it was found that the total charge obtained increased about 25 percent by bending it as shown. This is probably caused by an increased number of particles striking the wall of the tube due to centrifugal force on them. No further charge was obtained by either lengthening or shortening the tube.

The charge given by 5 grams of powder reaches the surprising value of 3.07×10^{-4} coulombs or 6.14×10^{-5} coulombs per gram. According to this value only 16 grams would have to be blown per second to get a current of 1 milliampere. Altogether 20 such runs were made, giving results which deviated at the most 11 percent from those given in Table I. None of the vagaries, such as reversal of sign, usually associated with frictional electricity were ever observed. A few runs made with lower air velocities gave much lower results, ranging from 3.02×10^{-5} coulombs per gram for the lowest air velocity capable of carrying the dust out of the cylinder and up. This is believed to be due to the cohering of the particles of the powder. The individual particles are approximately 10^{-4} cm in diameter and smaller, but they cling together forming larger aggregates which are blown apart by the air stream, the more completely the higher the velocity. It seems likely that larger charges per gram might be obtained by using higher air velocities, but this point could not be proved because the air pressure available was limited to two atmospheres.

The diatomaceous earth used to obtain the above results contained 12 percent of adsorbed water. No difference resulted by using the powder dried at 300°C for 1 hour.

5-7 Continued.

A small scale model of a high voltage generator using diatomaceous earth blown by air was constructed as shown in Fig. 4. In this figure, *A* represents an insulated sphere of spun copper 20 cm in diameter, within which 8 copper tubes *B*, 0.5 cm inside diameter, were mounted. Compressed air introduced at *C* carried along with it diatomaceous earth introduced by a small screw conveyor *D* from a reservoir and blew it through the copper tubes. The charged powder left through a short length of glass tube and escaped into the air.

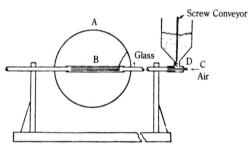

Fig. 4.—Experimental high voltage generator.

The potential of the sphere was estimated from the distance between it and a similar grounded sphere placed at such a distance from it that a thin spark jumped the air gap between them. The maximum potential reached seemed to be limited by a corona discharge from the sharp edges around the two openings in the sphere. The potential was estimated to be 260 kilovolts. A current of 8×10^{-5} amperes could be drawn from the sphere at this voltage by connecting it to ground through a glass tube filled with water and in series with a microammeter. The amount of powder introduced per second by the screw conveyor was 1.5 grams per sec.

In conclusion the writer wishes to thank Professor Millikan for very kindly placing the facilities of the California Institute of Technology at his disposal.

5-7 Continued.

of mica and shellac binders. U.S. patent 821,902 by Henry Todd (1906) mentions discs made from fibrous material and treated in a bath of molten sulphur. U.S. patent 1,109,205 by James Dempster (1914) covers discs coated with silica powders to increase surface area and reduce moisture condensation.

Closely related, but not patented, was a description published in 1910. Someone found that if the shellac varnish used by Wimshurst to coat his glass discs is first treated to increase its conductivity, the generator voltages and currents also increase.

The critical process is as follows: Into a clear glass bottle, pour your batch of white shellac varnish. Chop up pure copper wire of small diameter and drop it into the varnish. The varnish, being acidic, attacks the copper. Store this varnish in a warm, dark, dry place for about a week, stirring occasionally. When held up to sunlight, the varnish should show a light green color. At this point, remove all copper to stop the transformation. When allowed to turn dark green, the high copper content makes the conductivity become too great.

Coat the glass discs by using Wimshurst's method. Here we see Karl Winter's principle of semiconductor materials showing itself again. The area of disc coatings is the least-explored avenue in this science.

Precision construction techniques

In a more sophisticated direction, Noel Felici (in the 1950s) showed the importance of precision machined rotors and stationary inductors on his electrostatic generator. The traditional influence machines were made with clearances ranging from 0.1 to 0.2 inch between the neutralizers and discs. Felici reduced the working clearance in his cylindrical machines to 0.01 inch, thus permitting a stronger electric field for rotor charging. Of course, in a double-disc generator, the distance between the two discs must also be kept small. These distances are difficult to obtain with thin plastic discs but somewhat easier to achieve with glass plates—which remain flat and have good temperature stability.

Precision construction with small clearances boosts efficiency in electrostatic generators just as it does in electromagnetic motors and generators. The magnetic flux circuit requires a small gap between the rotor and stator to maintain a strong magnetic field and to conserve the energy needed to produce that field.

Liquid, gas, and vacuum chambers

Another method for increasing the electric field strength across generator parts, extensively explored from the 1930s to the 1960s, is to completely enclose the generator in a chamber for experimenting with compressed gases, dielectric liquids, or vacuum. Even though the use of compressed gases in electrostatics was recommended as early as 1886 by the German scientist Hempel, his discovery was ignored until Robert Van de Graaff revived the idea in the early 1930s!

The use of dry compressed air, carbon dioxide, nitrogen, or hydrogen can greatly improve efficiency. An electrostatic generator in an environment of pure hydrogen at 15 to 20 atmospheres with an efficiency of 90 percent has been built. Even though this experiment occurred in the 1950s, there is still a widespread misconception that this type of generator is merely an inefficient novelty or toy.

Although this method of improving efficiency is quite exotic for those with limited shop facilities, one simple generator was insulated with compressed air. This invention received British patent number 22,731 in 1900 and the inventor was a Mr. Tudsbury. This generator was a small Wimshurst unit enclosed in an airtight metal case. The compression seal and packing nut provided for the hand crank extended through the case wall.

Tudsbury claimed that a generator with 8-inch-diameter plates produced sparks 2¼ inches long at normal air pressure, 6 inches long at 15 psi, and 9 inches long at 30 psi!

Again, the compressed air must be dried. It is also a good practice to include a drying agent, such as silica gel, within the chamber. Now that acrylic and Lexan plastics are on the market, the chamber could be made transparent for inspection purposes. If you are working along this line, you should design the unit for easy accessibility and generator adjustment.

You can see from this discussion that many avenues are open for generator design. The results are limited only by your imagination. If even a small portion of the creative talent and money now invested in electromagnetic and semiconductor technology was applied to electrostatic generators, entirely new technologies could open up in the near future.

6
CHAPTER

Theories of generator operation

In this section, two samples of the orthodox, or textbook, explanations of how influence or induction generators work are included. It is fair to say that there are as many theories as there are inventors of original generator designs.

From the 1880s through the 1920s, heated debates appeared in the electricity journals concerning whose idea worked the best. The issue still has not been resolved; that is, no single theory explains all the properties that characterize these generators.

Because this is mainly a how-to book, I have included only the two most popular theories and limit the discussion to Wimshurst's disc-type generator (Figs. 6-1 and 6-2).

The first theory explains the mechanism as one of *influence*, or inductive action by one plate and its sectors on the opposite plate. The second theory avoids reliance on induction and is novel in its use of relative motion between charged bodies. There, the electric field is seen as stretching or relaxing, like a rubber band, between the plates and the neutralizing rods.

Presently, the influence machine is thought of as a continuously variable condenser; the discs form plates. In the relationship $V = Q/C$, V is the voltage, Q is the charge in coulombs, and C is the capacitance in farads. If we can decrease the capacitance value when a fixed charge is given to a capacitor with movable plates, the voltage or potential difference across the plates will increase.

Collector combs and Leyden jars can be seen as accessories that are not needed for the generator to begin charging. Wimshurst did make a simple machine, with only the two plates and two pairs of neutralizer rods, which worked well. The metal sectors are not needed, and the generator is more efficient without them. The sectors mainly provide the convenience of self-starting.

Sectorless influence machines, such as I prefer to make, can in a sense be viewed as having two plates with an "infinite" number of tiny sectors. The surface particles, which compose the discs, become the charge carriers—the charge leakage is thereby reduced.

Machine with oppositely rotating Plates.—We now come to the type of machine which has been most recently introduced by Mr. Wimshurst, and which is more especially associated with his name.

It was first described in January 1883, and, as may be seen from the illustration (Fig. 62), differs essentially in construction from either of the machines which preceded it. It also differs greatly in its behaviour, for it is self-exciting, and will discharge its torrents of electricity under atmospheric conditions which are fatal to the working of the other forms of electrical machines. Moreover, the direction of the current will not change when the machine is at work; nor will the excitement die away when the terminals are opened beyond the limit of the sparking length.

Following the plan which we have already adopted, in dealing with other machines, we shall explain the action of the Wimshurst machine by means of a diagram (Fig. 61), before describing the different forms of it which have been constructed.

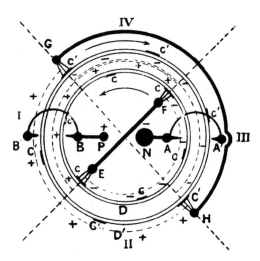

In the diagram, the oppositely rotating discs are represented by Bertin's method, as sections of cylinders D and D^1. The inner circle D represents the front disc, the outer circle D^1 the back disc, and each rotates in a

6-1 Diagram of Wimshurst machine.

direction shown by the adjacent arrow. On each disc there is fixed a series of metallic carriers, denoted by the letters C and C^1. A neutralising rod E F is fixed on the disc D, so as to connect diametrically opposite carriers as they pass and touch the contact brushes fixed on its extreme ends; and a similar neutralising rod G H is mounted on the face of the disc D^1, with its contact brushes in a diameter at right angles to E F. The electrodes B P and A N have discharging knobs P and N at one end, and collecting combs B B^1 and A A^1 at the other. The collecting combs B B^1 stand facing each other, one in front of each disc, and are metallically connected by a bent rod passing round the edges of the discs; A and A^1 are fitted up and connected in a precisely similar manner.

This machine, like all those having metallic carriers, is self-exciting, that is to say, it requires no charge of electricity imparted to it from an external source to set it in action. The action of the machine does not depend on the presence of the collecting combs; for the charges on the surfaces of the plates are produced though the collecting combs are removed. In considering the development of electricity in the machine, the collecting combs and the electrode circuit may be left out of account. It is not yet certainly known how the initial charge is produced which leads to the starting of self-exciting machines, but probably it is due to the fact that there are no two places in the atmosphere at exactly the same potential at any given moment. As a consequence of this, we shall suppose that one half G B^1 H of the disc D^1 has a small positive charge, and the other half H A^1 G has a small negative charge. Under the influence of the positive charge on D^1, each carrier C on D, when in contact with the brush E, will receive a negative charge, which will be carried round by the motion of the disc till every carrier on the half E A F of the disc has a negative charge. In the same way, the influence of the negative charge on D^1 will impart a positive charge to each carrier C as it passes under the brush F, by which operation the half F B E of the disc D will be coated with positive electricity. The two halves into which the disc is divided by its neutralising rod have thus far acted

6-1 Continued.

as field plates, to induce charges on the two halves into which the disc D is divided by its neutralising rod. The two halves of D now in turn act as field plates to induce charges on the carriers C^1 of the disc D^1, and a little consideration will show that, owing to the motion of this latter disc, the charges originally upon it will be increased by this action. The discs will continually react upon each other in the manner just described, and raise the potential of each other's charges according to the compound interest law, till the limit fixed by leakage is reached. The final distribution reached, is shown by the dotted lines, and + and – signs on the diagram. If we consider the discs divided into four quadrants (I, II, III, IV), by the two neutralising rods, it will be found that the charges on the two discs in quadrants II and IV are opposite, and, therefore, attract each other, but on the two quadrants (I and III) they are the same, and therefore repel each other. The collecting combs are placed opposite the middle of the quadrants (I and III) to draw off the self-repelling electricities. This, as shown in the diagram, imparts a positive charge to the discharging knob P, and a negative charge to the discharging knob N, and if these knobs are not too far apart, sparks will pass between them. As the action of the machine takes place without the collecting combs, it is not necessary that they should be in contact in order to start the machine. It is for the same reason impossible to reverse the action of this machine by separating the discharging knobs beyond the striking distance.

6-1 Continued.

Generally speaking, there are a number of generator properties, which an adequate theory would need, to explain the following characteristics:

- The source of the initial charges in self-starting generators with sectors. Both contact potential and cosmic rays have been invoked.
- The buildup of potential and what determines the upper voltage limit, other than corona leakage.
- Sudden polarity reversal, which was mostly a problem with the Holtz and Varley generators. Wimshurst machines generally have a stable voltage output.
- The importance of the disc surface—its microstructure, semiconducting or insulating properties, and surface charge distribution.
- The role of the surrounding medium, whether it is dry air, compressed gas, dielectric liquid, or vacuum. Why is pure compressed hydrogen gas very effective in spite of its low dielectric strength?

- The generator power sometimes spontaneously dies out when the discharge terminals are separated beyond the maximum striking distance.
- The sensitiveness of generator performance to neutralizer system design and its circuit resistivity.
- The disc material decomposes with time. In the case of acrylic plastic, the plates become coated with fine white dust after extended periods of operation. The conventional explanation is that generator discs are decomposed by ozone gas.

These are the major characteristics that have appeared after many hours of experimentation. Of fundamental importance is the very nature of *electrification*, whether by friction, impact, contact, or induction. I will discuss the subject of electrification later in chapter 10.

---

647. *New Theory of the Wimshurst Machine.* **F. V. Dwelshauvers-Dery.** (Deutsch. Phys. Gesell., Verh. 4. 13. pp. 276–277, Sept. 23, 1902. Paper read before the 74 Naturforscherversammlung at Carlsbad.)—At the request of de Heen the author contributed the following results: If an insulated body AB is moved rapidly towards a charged body C it becomes similarly charged (see Fig. 1). This phenomenon cannot be due to induction, since the charges at A and B are similar. If the motion is in the opposite direction so that the conductors are separated the sign of the charge in AB is reversed. These experimental results suggest the following explanation of the Wimshurst machine, in which the second plate is indicated by dotted lines and dashed letters. Suppose that there is a trace of positive charge in C (Fig. 8). The elements of the sector bOC are approaching C ; hence they become positively charged, and the charge in C is increased. In the sector

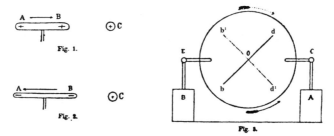

COd the reverse takes place, though the change is numerically less since the surface is less. The elements of the sector dOE become negatively charged since it is receding from C, hence E becomes negatively charged. A trace of electricity in either C or E thus determines the poles of the machine. In the above only one plate has been dealt with. The machine does give charges when only one plate is used, but they are so small that an electroscope is required for their detection. When both plates are used the second acts like the first, the positively charged sector b′OC increases the charge in C, and the negatively charged sector increases the charge in E. The most important action, however, is that of the one plate on the other. In the angle dOd′, for instance, the positively charged elements are approaching one another, and hence become more and more charged. This action is greatest when the distance between the plates is as small as possible. J. E.-M.

6-2 An alternative theory. *Science Abstracts, A, 1903*

7
CHAPTER

Van de Graaff generators

Prehistory

Leather Electrical Machine—The Rev. Theo. Dury, under date of December 17, 1838, at Keighly Rectory, Yorkshire, England, communicates the following fact to Dr. Faraday. He speaks of what he calls an extraordinary electrifying machine, which is no other than a leather strap which connects two drums in a large worsted mill in the town of Keighly. "The dimensions and particulars of the strap are as follows:—It is in length, 24 feet; breadth, 6 inches; thickness, ⅛ inch; it makes 100 revolutions in a minute. The drums, over which it passes at both ends, are 2 feet in diameter, made of wood, fastened to iron hoops, and turning on iron axles; these drums are placed at ten feet distance from each other, and the strap crosses in the middle between the drums, where there is some friction; the strap forming a figure of eight. There is no metal in connection with the strap, but it is oiled. If you present your knuckle to the strap above the point of crossing, brushes of electrical light are given off in abundance; and when the points of a prime conductor are held near the strap, most pungent sparks are given off to a knuckle at about two inches. "I charged," says Mr. D. "a Leyden jar of considerable size in a few seconds by presenting it to the prime conductor. The gentleman who told me of this curious strap has frequently charged his electrical battery in a very short time from it; and he informed me, that it is always the same, generating electricity from morning to night, without any abatement or alteration. If this strap had the advantage of silk flaps and a little amalgam, it would rival the machine in the lecture room in Albemarle-street.— *Silliman's American Journal of Science and Arts.*[1]

Although modern continuous belt-driven electrostatic generators are said to originate with Robert J. Van de Graaff in 1929, there have been a number of similar

---

1 From "Notes and Notices" section of *Mechanics Magazine* (London), vol. 32, January 1840, pp. 255–6.

generators described much earlier. Von Busch in 1893 showed his device having two pulleys and a horizontal belt with a charge collector comb and insulated sphere. Earlier still in 1785, Rouland invented one using a continuous silk ribbon running between two horizontal pulleys with a collector tube at midpoint.

Ever since the invention of flat belts and wooden pulleys, the potential for magnified electrical effects was possible. But it was unlikely that the historical records would be preserved down through the centuries. It is also a sad fact that superstitious fear of the unknown in earlier ages and later, in the industrial age, the attitude of treating electrostatic phenomena as nuisances hindering a production process postponed the development of this branch of science.

In hindsight, today's electrical experimenter should be open to observing small subtle effects in Nature and in their research since these may blossom into a whole new area of inquiry.

Dr. Robert J. Van de Graaff had been considering high-voltage generators for use in acceleration of charged particles for exploring the atom. In 1929 he cobbled together his first invention, the "tin can generator," using a pure silk ribbon driven by a small motor with charges collected in a tin can terminal. It developed about 80,000 volts, the limit fixed by the edges of the can. On February 12, 1935, Van de Graaff was granted U.S. patent 1,991,236 and potentials up to 10 million volts were produced with his design methods.

Most Van de Graaff generators used for demonstrations in high school and college physics classes produce a potential difference by "self-excitation," whereas those used for research have a separate power supply to spray charges onto the belt and are "charge-spray excited" (Fig. 7-1).

Almost all descriptions of the self-excitation principle say charges are produced by "friction," i.e., rubbing or slippage between belt and pulley, but this is not an accurate description. Electrification by rubbing or frictional slippage is called "triboelectrification" and is often experienced by sliding across a vinyl seat cover in the car and touching the metal door handle. Other common examples are moving across a carpet in winter and touching a door knob or separating warm clothes taken from the electric dryer.

Yet if I put a mark on the Van de Graaff belt and an adjacent mark on the pulley and run the generator for a minute, I see very little slippage between the two. This means that the real explanation of how self-excited Van de Graaff generators produce a potential difference is not by rubbing or slippage but simply by *contact* between the belt and pulley surfaces. The technical term is called *electrification by contact*, which is discussed at the end of this chapter.

Theory and construction

In Fig. 7-2, we see the utter simplicity of the generator's design, which is the reason they may be purchased at low cost and are still popular in the classroom. In operation, electrons are removed from the belt at the bottom plastic pulley where they jump onto the charge-spray screen or "comb" and to earth via the grounded motor and base. The portion of the belt moving upward therefore has a net positive charge.

7-1
Self-excited Van de Graaff
(Science First, Buffalo, NY)
14-inch terminal; 500,000 volts.

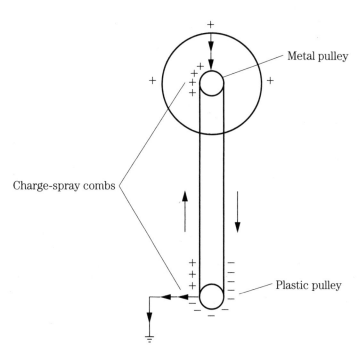

7-2 Internal construction (positive terminal).

As the belt passes over the top metal pulley, free electrons from the top terminal are sucked off the top screen or comb and onto the electron-deficient belt and then down to the lower plastic pulley where the cycle is repeated thousands of times per minute.

In essence, the generator is like a high-pressure (voltage) and low-volume (current) pump. In one to two seconds, the potential of the upper terminal will rise to a maximum *positive* value of several hundred thousand volts with respect to the base terminal shroud. The top terminal is positive because of the placement of the plastic and metal pulleys. To reverse polarity you simply use a metal pulley on the motor at the base end and put a plastic pulley at the top end (Fig. 7-3), resulting in a negatively charged terminal.

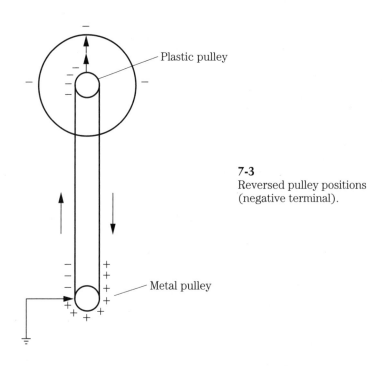

7-3
Reversed pulley positions
(negative terminal).

Two Van de Graaffs, one positive and one negative, will develop spark lengths twice as long because the potential difference (voltage) is now double between the two top terminals.

Modifications and improvements in design

As there are already many Van de Graaff generators in school physics labs throughout the country and they are readily available through science equipment supply catalogs, it is worthwhile discussing improvements in their performance. An excellent classic book that covers the subject well is the *Amateur Scientist* by C.L. Stong (1960). In this work Mr. Franklin B. Lee, who developed the Morris and Lee scientific instrument company, gives some rules to follow in this matter. One rule of thumb is that about 50 square inches of belt per second passing over the pulleys can produce 1 microampere of current in self-excited generators. The theoretical current for a belt ¾ inch wide on pulleys ⅞ inch diameter rotating at 3500 rpm is about 5 microamps. Because of losses, actual current will be 2½ to 3 microamps.

To measure the output, connect one terminal of a microammeter (0–50 microamps range) with a jumper cable to the wire upper comb assembly. The generator terminal must be removed for this. Connect the other meter terminal to a known ground at the base. The current output of self-excited Van de Graaffs is improved by increasing both belt speed and belt width. The belt should be slightly less than the pulley width and edges of a centered belt at least ⅛ inch from the inner column wall. Shim pulley at top with a cardboard strip under the axle to properly center belt. Belt speed is limited by increasing vibrations in the belt, belt tension, and by pulley bearing wear at high speed. It normally is 3000 to 5000 rpm, or less than 100 feet per second.

Check both spray combs, also called brushes, for their design and placement. Aluminum screen wire tends to fray, so use superglue to set the edges and cut points straight across. Or, replace screen with a sheet of 0.002-inch brass or steel shimstock cut across with pinking shears to give a row of points. The width of comb points should be about ⅛ inch less than the width of the belt. Remember that spray combs never should touch the belt and are best set at ¹⁄₃₂ inch from the belt surface with points evenly spaced from it (Fig. 7-4). Copper wire leads are soldered to the shimstock for connections. Check both combs in the dark and note the blue glow is uniform across the belt but does not extend to the underside of the belt.

7-4
Serrated charge spray comb/
0.002-inch shimstock.

Support column

Most columns are made from grey or white PVC pipe, but many teachers prefer clear columns so that the belt can be seen. This also allows for dust leakage to be easily spotted in the dark. Acrylic tubes are not serviceable because they can develop hairline stress cracks at both ends due to sparks. But Tenite Butyrate tubing used for mailing tubes is a good inexpensive material. It also has a hard surface for easy cleaning. Because it is thin-wall, I strengthen both ends by slipping over short PVC collars cemented on with silicone caulk.

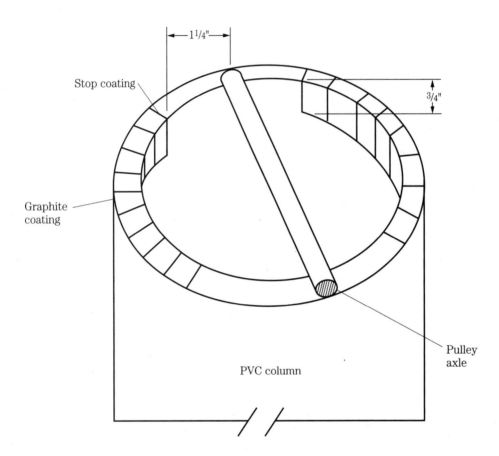

7-5 Graphite coating at ends of column.

Check both ends of the supplied column for graphite-blackened areas (Fig. 7-5). This blackened area helps remove extraneous charges thrown off by the belt, which inhibit generator performance. But the coating must be confined to the ends of the column as a ring, should not extend into the column more than ½ to ¾ inch, *and* should not come closer than 1¼ inches to the pulley axle or motor shaft. If it does, remove the excess by scraping with a utility knife. Do not use sandpaper or acetone as this will contaminate the rest of the surface. Too much graphite coating inside the column short circuits the generator and prevents longer sparks.

Next, check inside the column with a flashlight for streaks of grease or dirt, and clean with an acetone-dipped rag taped onto a yardstick. The inner wall should shine and be very clean and smooth. Check metal column support assembly and clamps at the base for sharp edges and smooth up or replace with well-round parts. Any metal screws passing through the column wall at the base should be replaced with nylon machine screws either #8-32 or #10-24 size. Any sharp metal points at the column base will prevent the full potential from being reached.

Base housing

The best design for the base housing cover is an aluminum shroud, hemispherical in shape just like the bottom half of the high-voltage terminal (Fig. 7-6). The shroud should be especially smooth at the re-entrant hole where the column passes through. The edge of the hole should be turned slightly inward (Fig. 7-7) for which I

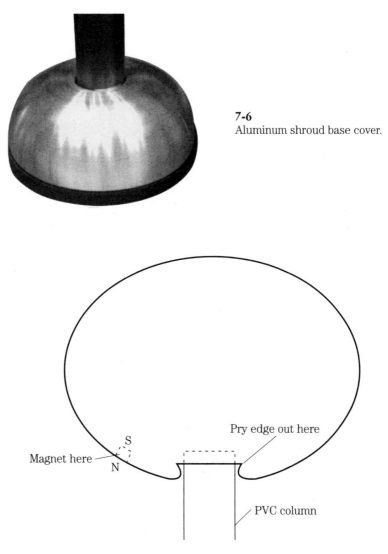

7-6
Aluminum shroud base cover.

7-7 Re-entrant hole edge. (Base and terminal.)

Magnet here

S
N

Pry edge out here

PVC column

fashion a tool from a strip of steel as shown (Fig. 7-8). It progressively pries the edge inward and this edge is smoothed with 120-grit sandpaper and then polished. This again will reduce leakage to the base. Any sharp corners in the base housing prevent uniform electric field distribution as can be seen in the dark. The metal shroud should be grounded to the motor frame and also to the electric outlet ground.

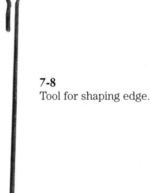

7-8
Tool for shaping edge.

High-voltage terminal

Perhaps the most important part of the Van de Graaff generator, which determines maximum voltage, is the top terminal. It is usually spherical or oblated in shape comprised of two spun aluminum sections with a horizontal seam. The seam and the edge of the re-entrant hole must be very smooth to the eye and to touch. A single sliver of aluminum can reduce voltage greatly. Remove excess solder and smooth edges with 400-grit sandpaper (garnet or emery) and follow with crocus cloth, which is a fine polishing cloth. I run a pin along the seam to dislodge small slivers. **Never use steel wool to polish any part of a high-voltage generator**, as a single strand becomes a major leakage source.

Use the edge turning tool (Fig. 7-8) to pry the edge of the re-entrant hole inward and smooth and round the edge as before. Check for pinpoints of light in the dark. Spin the terminal with discharge tongs with generator running and note if leaking points move also. If the point of light stays fixed, move the generator away from other objects in the room. Remember that a Van de Graaff rated at 500,000 volts can leak charges to any non-plastic object up to 4 feet away from the terminal. Even pegboard and plaster are seen as ground.

Another rule of thumb for those designing generators is that the theoretical voltage of a Van de Graaff terminal is equal to 76,000 volts per inch times the smallest terminal radius in inches. So a 7-inch-diameter terminal can reach 76,000 times 3.5 inches or 266,000 volts and a terminal 14 inches diameter can reach 532,000 volts. These values are determined by the insulating property of dry air called the voltage gradient. Its maximum is 30,000 volts per centimeter; beyond this, electrical breakdown occurs. The actual potential reached will be 5 to 15% less because of losses at the hole and to the base. Polishing the terminal adds very little to its efficiency but lint, dust, and hair can reduce its efficiency up to 40%! Cleanliness is next to godliness in electrostatics.

Pulleys and axle bearings

The plastic pulley needed for Van de Graaff's generator should have good dielectric properties, have a smooth surface that is easily cleaned, and not absorb moisture. Some designers have mistakenly put fur on plastic pulleys, not understanding that charge generation is by contact rather than slippage. This should be removed and a fresh surface turned on the lathe or the pulley replaced. The plastics most often used are white polyethylene or Teflon. These need not be solid pulleys and for larger sizes may be sleeves press-fit onto wood cores. Sleeves should be thick enough to crown the surface so that the belt will ride in the center.

The conducting pulley is usually aluminum, but hardwood, micarta, and Bakelite can also be used. Plastic pulleys can be transformed into conducting pulleys by covering them with strips of adhesive aluminum sealing tape (used on air-conditioner ducts) as in Fig. 7-9. I use superglue if the tape doesn't stick well, then I press out bubbles and creases to get a smooth aluminum surface.

If a negative high-voltage terminal is desired, replace the top conducting pulley with a plastic one. Check to see if axle has bushings or ball bearings. The use of bushings is not recommended because of excessive noise and strain to the motor. I use two ball bearings, 0.250-inch hole and 0.625 inch o.d., and a bearing spacer of vinyl

7-9 Plastic pulley resurfaced with aluminum tape.

tubing with a brass or steel axle. The pulley is centered on the axle with two ¼ inch shaft collars as shown (Figs. 7-10 and 7-11). I have found that a negative terminal gives slightly longer sparks. The use of bearings prevents possible motor stalling and helps the teacher and students to converse and hear quieter leaking sparks.

7-10
Upper plastic pulley with bearings.

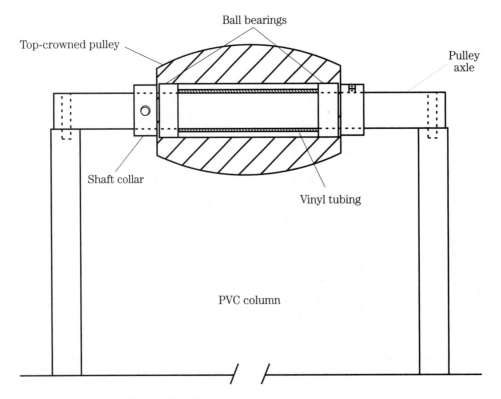

7-11 Internal view, pulley and bearings.

Belts

Replacement belts of neoprene are available at reasonable cost and can have a lifetime up to about 400 hours. Gum rubber belts have a much shorter lifetime. The material should be slightly stretchable, have a high dielectric constant and density, and be resistant to ozone. Neoprene has these properties. Latex rubber and neoprene belts can be cut from 0.025-inch-thick sheet and spliced. The splice is a smooth butt joint, not a lap, and is cut at a 45° angle with a razor blade. The two edges must be straight-cut so that edges make contact.

Place these edges near each other on a 2-inch-wide clear plastic block which has been covered with a strip of polyethylene sheet. Apply a thin bead of superglue **gel** along one of the edges of the belt and push the ends together, checking for good contact. Squeeze the joint slightly using a second sheet-covered block and two clamps (Fig. 7-12). You can see into the block sandwich to check this joint. Set aside for several hours. The sheet covers prevent the glue from sticking to the blocks.

7-12 Diagonal belt splice setup.

Trim the edges of the joint to an even width. Degrease belts with denatured alcohol, and after every 50 operating hours, clean with soap and water, then rinse with distilled water. Store dry belts in a Ziploc freezer bag. Remember to allow for belt stretch by cutting belts about 2 inches less than the actual circumference around the pulley set. For a Van de Graaff with a motor speed of 3000 rpm and belt width of 2 inches, you should get 10 to 12 microamps charging rate in dry air.

Motor drive

Replacement motors should be small and quiet and completely enclosed. It will be easier to add speed control using 12-Vdc permanent magnet motors. Motors with open windings should be avoided, since these hold moisture that is released into the column and spoils output. A small light at the base of the column can help heat the column and drive out moisture. Figure 7-13 shows a Science First model #10-085 Van de Graaff modified as described to give up to 19-inch white-hot sparks having an apparent thickness of ⅛ to ¼ inch. This result is possible without any extra capacitor.

7-13 Modified Science First Van de Graaff model #10-085 negative terminal, 19-inch hot sparks.

Useful accessories

Goosenecks used for lamps can be joined to the base plate with a 90° pipe elbow and the aluminum shroud slotted at the joint as shown in Fig. 7-14. Terminal balls 2 to 4 inches are then attached to the gooseneck with brass lamp couplings. This extra terminal outlet can be adjusted for shorter spark lengths; the gooseneck is laid down horizontally for the full-length sparks.

7-14 Gooseneck to base attachment.

A simple plastic clamp (Fig. 7-15) can be positioned up or down on the column for supporting electrostatic tops and capacitors. Note that all fasteners must be nylon.

To attach steel spray points to the aluminum terminal, I attach rare earth magnets inside the lower part of the shell using superglue and mark an ink dot on the outside for location. Never drill holes in terminals or attach plugs for jacks, as this will ruin the performance through leakage. These magnets can support a nail or chain (Fig. 7-16) useful for experiments.

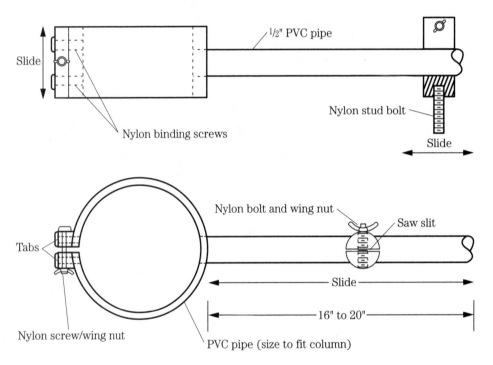

7-15 Experiment support clamp.

Triboelectrification
and contact electrification

One of the areas of solid-state physics long explored but least understood concerns how charge is produced between two solids when they are rubbed together and then separated. The amount of charge and the net polarity depends in part on what material is used and the condition of its surface. This frictional electricity, also called *triboelectricity*, has been studied for several centuries and a relationship among materials has been established, a general ordering called the triboelectric series. It lists solids that, when rubbed with another on the list, the one listed higher receives a positive charge and those below it a negative charge.

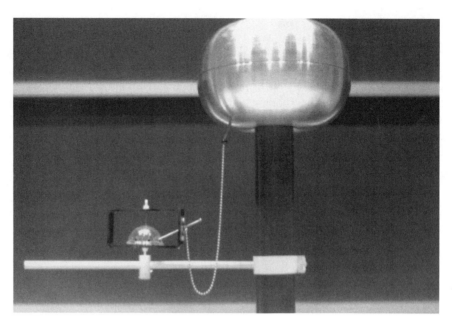

7-16 Internal magnet in terminal.

One such list follows:

Positive (+)
mica
wool
nylon
polished glass
lead
silk
aluminum, zinc
filter paper
cotton
dry wood
unpolished glass
Lucite/Perspex
paraffin wax
polyvinyl chloride (PVC)
polystyrene
polyethylene
soft rubber
sulfur
hard rubber
Teflon
Saran wrap
Negative (−)

The first published series was by J.C. Wilke in 1757 and another one produced by P.E. Shaw in 1917 and another by D.J. Montgomery in 1959. But there was not complete agreement and no rule was established for determining the position of materials in the series. This caused a number of solid-state physicists to doubt the validity of any triboelectric series. Harold F. Richards published his work on impact electrification in 1920, showing that it is not the rubbing that is fundamental but rather intimate contact between two surfaces. When a hard rubber ball bounces on an insulated brass disc, the disc can reach a potential of nearly 200 volts! Richards then tried to relate contact electrification for insulators to contact potential in metals (called the voltaic series).

Even though the triboelectric series is most often used to reduce the potential differences as needed in the textile industry, grain elevators, and oil tankers, it can be useful in improving the design of frictional electrostatic generators. The mystery of how contact between surfaces produces a transfer of charges still remains a wide-open avenue for the electrical experimenter.

2
PART

Accessory instruments, experiments, and applications

In the second section of this book, I will describe some of the basic electrical devices that can be used for extensive exploration into the nature of electrification. Although the cost of the sectorless influence machine that I describe might range from $100 to $200, depending on the tools available in your shop, the electroscope, Leyden jars, and the electrophorus are much less expensive. These latter projects require only simple and inexpensive tools for their construction. However, do not assume, as most textbooks on electricity would lead you to believe, that these instruments are well understood in operational theory.

Following the electrophorus discussion, I will finish with some of the many unusual experiments in high-voltage electricity that are rarely mentioned in classroom physics books. You will see that there are a number of unexplored avenues awaiting the pioneer in electrical science.

8
CHAPTER

The electroscope

The *electroscope* is a device that determines the electrical condition of its surrounding atmosphere; specifically, it determines the presence of charges on nearby bodies or the polarity (+ or −) of each charge. The basic electroscopic instrument does not display an absolute numerical measure; instead a simple scale is provided, graduated in degrees. As in Fig. 8-1, a typical design consists of a metal rod, on top of which is mounted a ball or disc, and carrying at its lower end either a single strip of gold leaf and a metal plate or two strips of gold leaf. The rod and leaf are electrically well insulated from ground, and the delicate leaves are protected from air currents. To establish that a body is electrified, bring it close to the top terminal. If the leaves diverge, charge is present. Normally, a glass or plastic rod rubbed with a dry cloth is used to charge the scope.

Figure 8-2 shows some of the possible electrified states that the electroscope can exhibit. The description of electroscopic charging in Fig. 8-2, written in 1875, is the same explanation that would be given today in physics textbooks.

8-1
A commercial electroscope.
Physikalische Technik, vol. 2, J. Frick, 1907

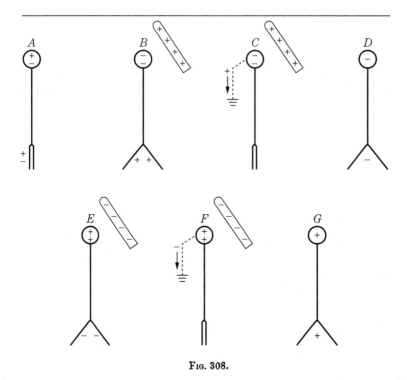

Fig. 308.

In the ordinary state the electroscope contains equal quantities of positive and negative electricity, as represented at A in fig. 308. When a body charged, say, with positive electricity, is brought near it, electrical separation is effected, the negative electricity being attracted to the knob, the positive electricity being repelled to the further extremity, that is, to the gold leaves. These are now electric, and being both charged with the same electricity, they repel each other, that is, being flexible, they *diverge*, as at B. When the inducing body is removed, the leaves drop again because the two separated electricities recombine, and the electroscope is again in the neutral state A. If the metal rod is touched with the finger, while the inducing body is still near and the two electricities therefore

8-2 Various charged states of electroscopes. *Introduction to Experimental Physics*, Adolf Weinhold, 1875

still separated, then, as shown at *C*, the free positive electricity of the gold leaves escapes through the finger and the leaves drop, while the negative electricity of the knob remains bound as long as the inducing body is near the instrument. Now, while the inducing body is still near, let the finger be removed first, and next the inducing body; then the negative electricity becomes free, and as it cannot now escape, it diffuses itself over the metal portion of the instrument, and the gold leaves diverge again, as in *D*.

The electroscope is now charged by induction with the opposite electricity to that of the inducing body, that is, negatively, if the inducing body was, for example, a positively charged glass rod, and positively if the inducing body had been a negatively charged stick of sealing-wax. The latter case is represented in figures *E*, *F*, and *G*, which correspond to figures *B*, *C*, and *D*, respectively.

An electroscope, *charged* with either electricity, may be used not only for deciding whether a body is in the electric state or not, but also with what kind of electricity the body is charged. When a neutral body is brought into the neighbourhood of a charged electroscope as shown in *A* and *B* of fig. 309, no appreciable change takes place in the divergence of the leaves. In reality the divergence does diminish in that case slightly, because the electroscope itself acts now like an inducing body, separating the electricities of the neutral body brought near it, and in consequence of mutual inductive action, a quantity of electricity of the opposite kind to that with which the instrument is charged, is repelled to the leaves, neutralising a portion of the original charge.

8-2 Continued.

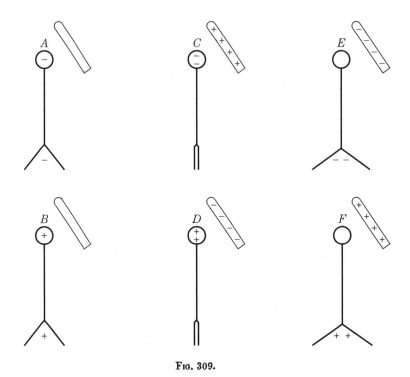

Fɪɢ. 309.

This action is, however, so slight, compared with the effects of a charged body, that no mistake can be made in the conclusions. If a body charged with opposite electricity is brought near to the instrument, as in *C* and *D*, the leaves drop because their charge is attracted to the knob. If a body charged with the same electricity is brought near, the charge in the knob is repelled to the leaves, and their divergence increases as in *E* and *F*.

8-2 Continued.

Later in the text are experiment results that suggest that the classical explanations are superficial and based on too few experimental tests. First, however, I will describe how you can build your own scope using common materials. A source is provided for the sulfur insulation for the leaf (see appendix).

The sensitive gold leaf is sold through picture-frame or craft stores in book form with about 25 leaves. Each leaf is about 5 inches square and separated by tissue paper. Each gold leaf is only about 0.1 micron thick (about four-millionths of an inch)!

The real gold leaf books are expensive, but if you are in a high school science class, you might share the cost of one book with several students. Otherwise, try composition gold leaf or aluminum leaf, which are quite inexpensive. Real gold leaf is much more sensitive because it is so thin and has a very low mass for its size.

Building electroscopes using tin boxes

Using a metal box with small viewing windows is preferred to glass bottles and flasks because the metal enclosure shields the sensitive leaves from extraneous electric fields so that even if a high charge exists on the outer box surface, the inner metal box surface remains uncharged or neutral. Such a metallic enclosure is called a *Faraday cylinder* or *Faraday shield.* The box also should be grounded for best results.

Fig. 8-3 shows two examples of fairly simple electroscopes. Once you have washed all those cookies down with gallons of Twinings tea, you can buckle down and begin building a tea-tin electroscope. Two rules are to keep the leaf and leaf support small for greater sensitivity and use an insulating material that holds a charge well and has low moisture absorption. Plastic bottle material, Teflon, and white styrene fashioned into thin washers constitute one approach, whereas casting sulfur bushings gives an insulator of highest resistance because sulfur is impervious to moisture.

8-3 Tea- and cookie-tin electroscopes.

The tea-tin scope design centers around use of the neck and cap removed from a large Ocean Spray fruit juice plastic bottle. Cut off the top 1 inch, which will include the cap, threaded neck, and neck flange (which seats the cap). Use a fine-tooth hacksaw for removal because utility knives are prone to slip off the hard surface. Once the top is removed, trim the edge with a utility or X-acto knife. Using Fig. 8-4 as a construction guide will help in fashioning the tea box. Two holes 1½ inches in diameter are cut into the front and back of the box; a hole saw will cause less distortion of the metal than sheet metal snips. Next, a 1⅝-inch hole is cut in the box lid and finished smooth with a

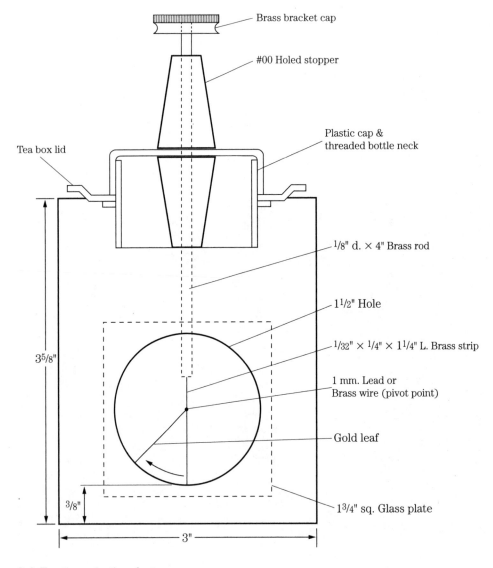

Brass bracket cap

#00 Holed stopper

Plastic cap &
threaded bottle neck

Tea box lid

⅛" d. × 4" Brass rod

1½" Hole

1/32" × 1/4" × 1 1/4" L. Brass strip

1 mm. Lead or
Brass wire (pivot point)

Gold leaf

3 5/8"

1 3/4" sq. Glass plate

3/8"

3"

8-4 Tea-tin projection electroscope.

file to let in the bottle neck. The plastic cap has a ³⁄₁₆-inch hole for the brass rod to slip through. Clean and paint the inside of the box flat black, and epoxy glue ⅛ × 1¾-inch square glass plates on the inside surface of the window holes; flashlight glass lens covers 1¾ inches diameter also work well.

The brass strip for mounting the gold leaf is cut, and all corners are rounded and polished. Approximately ¾ inch up from the bottom of this brass strip, solder or glue a horizontal section of wire or pencil lead to act as the pivot point and attachment point for the leaf. Solder the top of the strip to the end of the ⅛-inch brass rod; the leaf support is now complete. Next, slip on one of the #00 holed rubber stoppers shown in Fig. 8-4. This is followed with the cap and the upper stopper; the two stoppers give clamping action to the rod, which gives room for adjusting the rod up and down. Try to adjust the rod so that the pivot point for the leaf is centered in the windows. Note that the insulating value of this electroscope depends on the thin plastic cap rather than on the rubber stoppers. In general, the larger the volume of an insulator, the easier it permits charge leakage.

Mounting gold leaf

It requires much skill, patience, and time to work with gold leaf but, once mastered, allows the electrical experimenter to make very sensitive charge-detecting electroscopes. Remember that gold leaf is about ¹⁄₁₀ micron thick (one ten-millionth of a meter)! This is thin enough to transmit light.

I prefer to tape a 5 × 8 inch index card down on the table so that the long edge of the card overhangs the table about ¼ inch. Using a *new* single-edge razor blade, remove two tissue papers (with the gold leaf sandwiched between) and place them on the index card. Referring to Fig. 8-5, cut an index card as shown using the razor blade to make the rectangular opening and extend the cuts at the four corners. Lay this over the

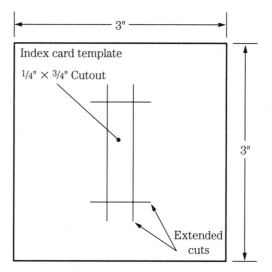

8-5 Gold leaf cutting template.

tissue and leaf sandwich. I prefer to pick up gold leaf with stainless steel tweezers tapered to rounded or flat ends about ⅛ inch wide with smooth, not serrated jaws. This reduces tearing and sticking problems with the leaf. Clean the tweezers and the brass strip with alcohol, and run a trace of saliva (or shellac thinned with alcohol) across the pivot point attached to the brass strip; this can be done with a toothpick. Be sure to keep the room draft-free and mask your nose and mouth to avoid blowing the leaf about. I advise you to stand over the template and hold it down with your fingers next to the rectangular slot. Then make the four *single*-stroke cuts with a *new* razor blade extending beyond the four corners so that tissue and leaf are separated at the corners.

As seen in Fig. 8-6, lay a small piece of the same tissue paper on the index card even with the overhanging edge, and tape the sides down. Using the tweezers, carefully lift off the top tissue over the rectangular section, and then hold the gold leaf by one end, slowly lifting it out. It should separate from the back tissue as well. Sometimes a small sewing needle helps separate the tissues. Note that even your body heat affects leaf movement. Transfer the leaf to the taped-down tissue paper, and place it so that one end is at the front overhanging edge of the card (see Fig. 8-6). Hold the leaf support as shown so that the shellacked wire is brought down slowly over the end of the gold leaf, touching it lightly, and then lift slowly. Care must be taken to join the brass strip and leaf parallel so that the leaf, when mounted, will deflect in a plane perpendicular to the plane of the support strip. If the brass strip is well polished and cleaned with alcohol,

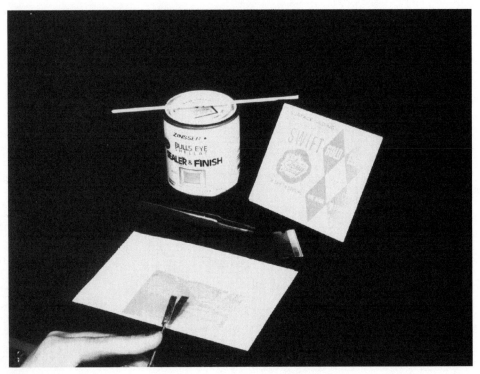

8-6 Mounting gold leaf to support.

the leaf will not stick when charged. In order to handle the leaf as little as possible, it is recommended that the leaf support, neck cap, and box lid already be assembled so that after the leaf is mounted, it can move directly into the box. Depending on relative humidity, the tea-tin electroscope will hold a sizable leaf deflection for about an hour; keeping the insulating cap clean and dust-free helps greatly. The top terminal of the scope may be a small brass disc; I chose a brass "bracket cap" (found in the lamp parts area of the hardware store). I solder a short piece of ⅛-inch i.d. brass tubing to the cap's bottom so that it slips over the ⅛-inch brass rod stem. This allows you to experiment with other terminal shapes, such as a small ball or point.

Figure 8-7 shows how to project the leaf shadow onto a white cardboard screen. You can experiment with a mini-Maglite flashlight that allows for focusing of the beam. This optical method greatly magnifies the image of the deflected leaf so that it can be seen by a class of students.

Building a cookie-tin electroscope

Get a cookie container such as is often used for packing fancy imported cookies. These metal containers measure about 3½ inches in depth and 7½ inches in diameter.

The completely assembled cookie-tin electroscope is shown in Fig. 8-8. This homemade inexpensive leaf electroscope with graduated scale can be used to measure high resistances, atmospheric electrical tensions, and radioactive samples. Figure 8-9 shows the interior of the electroscope.

8-7 Projecting leaf shadow onto screen.

8-8 The cookie-tin electroscope.

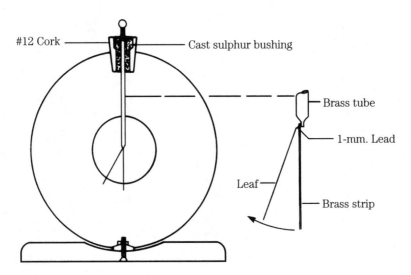

8-9 Construction sketch for the electroscope.

Directions for building the cookie-tin electroscope

Clean the cookie tin with soap and water, then dry. In the front and back, cut out a 2½-inch-diameter opening and remove the burrs for a smooth edge.

Into the top of the cylinder, cut out a 1-inch-diameter hole for the cork and smooth the edges. If you have a thin rubber grommet available, you can use it to line the hole; it looks more attractive.

Saw the wood base to measure ¾ × 2 × 8 inches and cut a saddle shape in the top surface for mounting the cylinder. Use a faucet beveled washer, flat-head machine screw, and nut to secure the cylinder to the base. Cut 3-inch-diameter windows from sheet mica or use glass lens covers (for flashlights) and glue them inside the 2½-inch-diameter holes, front and back. Put a thin film of glycerine on the outer surfaces of these two windows to make them conductive.

The leaf support is a thin wall brass tube ⅛ inch i.d. and 4½ inches long. Cut a strip of brass ¹⁄₃₂ × ¼ × 1⅝ inches, round its ends, and polish well to remove all sharp edges. Solder this strip on the flattened bottom of the tube, as shown. Glue a ¼-inch length of 1-mm-diameter pencil lead horizontally across the back of the brass strip to provide the point of support for the gold leaf.

Next, cut a #12 cork with a stepped through-hole, one size of which is ¾ inch in diameter and the other, ⅝ inch in diameter. Provide good clearance from the brass tube in the center.

Cover the bottom of the cork with adhesive aluminum foil, making sure it is well sealed around the bottom and side of the cork. Using a pointed tapered rod, pierce a ⅛-inch hole in the center of the tape and carefully enlarge it so that the brass stem (⅛-inch i.d.) just slides through. Find the desired position between the tube and the top of the cork by inserting the cork in the 1-inch hole of the cookie tin. The 1-mm lead (see Fig. 8-9) must be centered in the window, since it is the pivotal point for the gold leaf. When you find the best position, mark it.

Now, lightly hold the brass tube vertically in the vise and rest the finished cork on top of the vise jaws in the desired position. Note that the hole in the aluminum tape fits tightly around the brass stem and that the stem is concentric with the hole in the cork.

Place chunks of lump sulfur (brimstone) in a small porcelain crucible and cover with the lid. **Caution:** Sulfur should not be exposed to fire, since it burns readily. Heat the crucible very gently over a low flame—overheating destroys the insulating property of sulfur. The molten sulfur will be dark orange.

Remove the flame and pour the liquid sulfur slowly into your #12 cork. Fill until the sulfur is level with the cork's top. Allow the bushing to cool slowly and avoid causing vibrations while the casting sets. After about 5 minutes, the sulfur will be crystallized, and medium yellow in color. When it is hard, but still warm, peel off the foil seal around the cork's base. You now have a custom-cast bushing made of sulfur—one of the best insulators! After one day, the sulfur will assume a lemon-yellow color if it has not been overheated.

Sulfur holds a charge much better than Teflon and styrene plastics—two otherwise good insulators. For your own test purposes, when the leaf is charged to a deflection of 80 degrees (almost horizontal), it requires two hours to drop 15 degrees, which is the 65-degree position (temperature 64°F, relative humidity 45 percent).

Follow the instructions for mounting gold leaf given previously for the tea-tin electroscope. Slowly raise the leaf support system and cork, and position them in the 1-inch hole at top. Solder the two removable terminals— a ball and a 1-inch-diameter disc—to ⅛-inch-diameter brass rod; these terminals slide into the top of the brass tube.

Be careful not to overcharge the leaf and tear it. Its maximum deflection should be about 90 degrees (the horizontal position).

To protect the insulating value of your electroscope, cover the cork and terminal with a metal cap. This measure prevents dust from settling on the sulfur bushing.

The electroscope was especially useful at the turn of the century. Then, it was used to study radioactivity and the natural ionization of the air. Both Ernest Rutherford and C.T.R. Wilson studied and designed very sensitive gold-leaf electroscopes. Rutherford's book *Radio-activity* describes a modified electroscope (Fig. 8-10).

Rutherford calculated the current sensitivity of an electroscope with a 1-liter volume and a 4-cm-long gold leaf and rod support system. The electric capacity C is usually 1 electrostatic unit. If V is the decrease of potential of the leaf system in t seconds, the current i through the 1-liter volume of air is:

$$i = \frac{CV}{t}$$

Since the fall of potential in air averages 6 volts per hour, then:

$$i = \frac{1 \times 6}{3600 \times 300} = 5.6 \times 10^{-6} \; electrostatic \; units$$

Therefore, $i = 1.9 \times 10^{-15}$ amperes!

The amazing current sensitivity of such a simple scientific instrument clearly illustrates that you don't always need exotic and expensive equipment to do careful research. But, much practice and manipulative skill is needed to produce these delicate devices. C.T.R. Wilson developed the skill of cutting gold leaf to such a degree that he could produce leaf strips ¹⁄₁₀ mm in width! You must try cutting gold leaf to appreciate this feat.

A scale for registering leaf deflection is made with tracing paper, on which increments from 0 (vertical) to 90 degrees (horizontal) are marked. The paper scale is positioned behind the outside back window; a light shining behind the scale illuminates it for easy reading.

It is especially interesting to lecture to large classes with the electroscope placed between the condenser lens and compound objective lens of a projector or "magic lantern." The light source will project a shadow of the leaves onto a large screen marked with a scale. This way, the movement can be shown to many students at once (Fig. 8-11). Typically, the condenser lens has a focal length of 10 cm, and the objective lens (for focusing the image) has a focal length of 5 cm.

Electroscope anomalies

Most physics texts describe the electroscope as basically an "ionization chamber." This description is given because the charge leakage from the system is said to be the result of ions in the air within the electroscopic chamber. In 1901, C.T.R. Wilson claimed that the leaf discharge rate is the same during day and night; therefore, leakage is not caused by light. Sir William Crookes found that when electroscopes are

56. A modified form of the gold-leaf electroscope can be used to determine extraordinarily minute currents with accuracy, and can be employed in cases where a sensitive electrometer is unable to detect the current. A special type of electroscope has been used by Elster and Geitel, in their experiments on the natural ionization of the atmosphere. A very convenient type of electroscope to measure the current due to minute ionization of the gas is shown in Fig. 12.

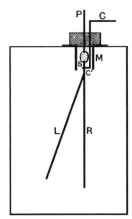

Fig. 12.

This type of instrument was first used by C. T. R. Wilson* in his experiments of the natural ionization of air in closed vessels. A brass cylindrical vessel is taken of about 1 litre capacity. The gold-leaf system, consisting of a narrow strip of gold-leaf L attached to a flat rod R, is insulated inside the vessel by the small sulphur bead or piece of amber S, supported from the rod P. In a dry atmosphere a clean sulphur bead or piece of amber is almost a perfect insulator. The system is charged by a light bent rod CC' passing through an ebonite cork. The rod C is connected to one terminal of a battery of small accumulators of 200 to 300 volts. If these are absent, the system can be charged by means of a rod of sealing-wax. The charging rod CC' is then removed from contact with the gold-leaf system. The rods P and C and the cylinder are then connected with earth.

The rate of movement of the gold-leaf is observed by a reading microscope through two holes in the cylinder, covered with thin mica. In cases where the natural ionization due to the enclosed air in the cylinder is to be measured accurately, it is advisable to enclose the supporting and charging rod and sulphur bead inside a small metal cylinder M connected to earth, so that only the charged gold-leaf system is exposed in the main volume of the air.

In an apparatus of this kind the small leakage over the sulphur bead can be eliminated almost completely by keeping the rod P charged to the average potential of the gold-leaf system during the observation. This method has been used with great success by C. T. R. Wilson (*loc. cit.*). Such refinements, however, are generally unnecessary, except in investigations of the natural ionization of gases at low pressures, when the conduction leak over the sulphur bead is comparable with the discharge due to the ionized gas.

---

* Wilson, *Proc. Roy. Soc.* Vol. 68, p. 152, 1901.

8-10 Rutherford's sensitive electroscope. *Radio-activity*, E. Rutherford, 1905

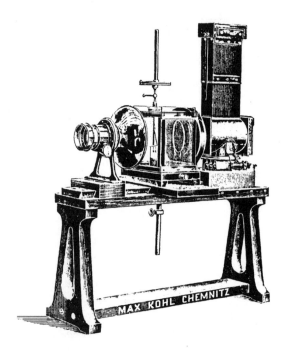

8-11
Condensing electroscope with lantern projector. *Physikalische Technik*, vol. 2, J. Frick, 1907

evacuated, leakage is greatly reduced, and in general, leakage is proportional to atmospheric pressure.

Unusual fluctuations in leaves, immediately after charging, are said to be caused by unequal heating of the air in the chamber from incident light beams. To determine if the orthodox explanation is justified, you can repeat Wilson's and Crookes' experiments.

One experiment that must be performed in clear, sunny weather, preferably outdoors, with a relative humidity of less than 20 percent, involves charging the scope to a deflection of about 50 degrees. After charging to this degree of deflection, remove the charging rod. Properly done, the leaf will increase its deflection spontaneously over the next few minutes by up to 20 degrees! It would appear that the energy is flowing uphill, from low to high, as though the charged scope is seeking a new level of energy equilibrium with its environment.

Now charge the electroscope by induction. The leaf then has a charge opposite to that of the charging rod. Slowly bring the rod from a distance toward the scope terminal. The leaf deflection decreases as you bring the rod closer. But as you draw the rod even closer, the leaf which is now at about zero deflection, begins to rise again. Finally, the leaves reach a high deflection when the rod is very close!

Is there an energy "node" around the instrument, with different properties on either side of the node? Can you think of a setup with more than one node about the electroscope?

Other anomalous electroscopic experiments vary the geometry of the materials making up the device and the types of materials themselves.

The quintessential, but unfortunately uncelebrated, electroscopic experiments were performed by the amateur experimenter, Dr. Gustave Le Bon of Belgium. His two books on physics, *The Evolution of Forces* (1908) and *The Evolution of Matter*

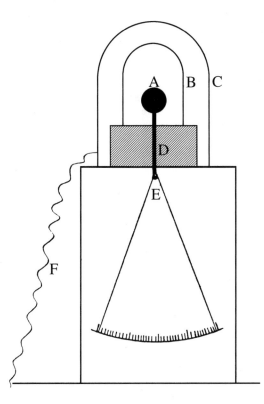

8-12
A screened electroscope. *The Evolution of Matter*, Gustave Le Bon, 1907

(1910), caused an uproar in scientific circles. I assume the uproar was because the works were so original and thought-provoking.

Figure 8-12 shows the electroscopic ball surrounded by three concentric metallic screens of varying thicknesses resting on an insulated plate. Under the influence of the charged horizontal rod above, the leaves diverge. Conventional theory says that electric forces should not penetrate the screens, since the screens act as Faraday cages.

Figure 8-13 shows three electroscopes, each with a different terminal shape over which is supported a negatively charged rod. Especially unusual is that the charged rod is kept positioned for some time, up to 12 minutes, in order for the leaves to be affected. When the rods are finally removed from the electroscopes, each will have a different electrified state—neutral, positive, or negative. Why is time an important factor?

In another experiment, a radioactive substance in a metal capsule is placed on the plate of the electroscope, and a metallic plate, such as aluminum, is positioned above on a stand. You might not expect the metal blade to influence the discharge rate, if the increased leakage is caused by radioactive ionization. The type of metal plate and its condition are also important. The same peculiar result was independently verified by physicists William Ramsay and W.S. Lazarus-Barlow about 1905.

The last of Le Bon's ingenious experiments I mention concerns electroscopic tests comparing the dissociation of metals by sunlight and radioactive substances (Fig. 8-14). Because Le Bon's experimental conclusions were so disturbing and seemed so unorthodox to the scientists of his day and because he did not happen to possess "credentials" as a physicist, his contributions were not acknowledged in

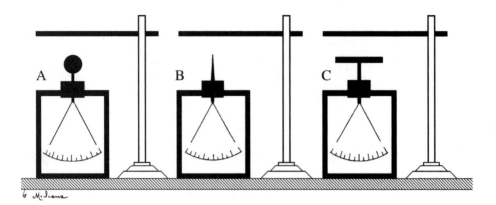

8-13 Effect of electroscope terminal shapes. *The Evolution of Forces*, Gustave Le Bon, 1908

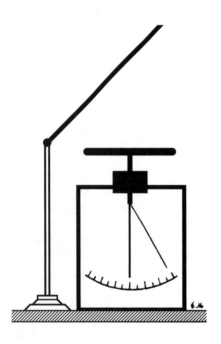

8-14
Influence of sunlight on metals.
Evolution of Matter

Fig. 14.—*Comparison of the dissociation of spontaneously radio-active bodies and of metals under the influence of light.*—A tin mirror prepared as described in the text and a screen of the same size coated with oxide of thorium or of uranium are used alternately. The dissociation of the atoms of the tin under the influence of light is forty times more rapid than that of the radio-active bodies just mentioned.

most journals. He was probably the first scientist to recognize and announce the universal and general dissociation of all matter. Many other physicists went on to build careers and receive awards for work that had beginnings in the research of this great scientific experimenter. Such is the true history of innovation!

Research avenues

In addition to using electroscopes to study the nature of electrification and the resistivity of insulators and semiconductors, you will soon realize that the instrument is very sensitive to weather changes.

Maximum leaf deflection and slow leakage occur when the sky is bright and clear, and the humidity is low. Minimum deflection and fast decay occur during humid, overcast days, in heavy fogs, and after heavy rains.

Several questions might occur at this point: How does the scope behave before and after thunderstorms? How is leaf deflection related to relative humidity? How is leaf deflection related to barometric pressure and temperature?

By placing a well-insulated bare wire horizontal to the ground and joining it to your scope, you can detect a buildup of charge, especially in clear sunny weather. How does the charge vary with the height above the ground? Does the type of metal make a difference? Is the sun radiating charged particles that electrify the wire?

Do electroscopes charge spontaneously during earthquakes? They have been shown to do so. An excellent book, which devotes a chapter to this subject, is titled *When the Snakes Awake—Animals and Earthquake Prediction* (1982) by Helmut Tributsch.

The connection between earthquakes and electrometers was first mentioned in 1799 in Venezuela. The world traveler, Alexander Von Humboldt, wrote that his electrometer oscillated oddly just before two earth tremors. Later, in 1808 the Italian scientist, Vassalli-Eandi, found the atmospheric electricity too large to measure during a tremor.

The weather during a quake is often described as being very still and oppressively heavy. Perhaps the piezoelectric or triboelectric effects cause an increase in charge.

How does the electroscope behave before the onset of a tornado? The electrical properties of tornadoes have been carefully documented in *Lightning, Auroras and Nocturnal Lights* (1982) by William Corliss. Homemade electric tornadoes are covered in chapter 19.

I have devoted considerable space to the electroscope in this book because its importance to basic research is seldom mentioned in physics texts. Its simple appearance is quite deceptive. Although electronic circuits are now available for measuring charges, they cannot deliver the useful information given from the study of leaf motions and the rates of leaf fall.

When all the anomalous experiments—usually omitted from classroom discussions—are taken into account, they point toward a need for a simpler, less-contrived explanation. The *mechanistic theory*, which relies on static forces, seems incomplete; what is needed is a *dynamic theory*.

Remember, when designing electroscopes, to surround the leaf system with a grounded metallic shield and to keep the dimensions of the leaf, leaf-support, and terminal small for maximum sensitivity to a given charge.

9
CHAPTER

The Leyden jar condenser

The *Leyden jar* is essentially a flat plate capacitor; its metallic coatings and dielectric are rolled up in cylindrical form. Figure 9-1 shows one of the earlier forms of the jar in use during the late 1700s.

The Leyden jar is recorded as being first developed by Ewald von Kleist in 1745. In 1746, Pieter van Musschenbroeck of Leyden, Holland, further experimented with the invention. The jar's original form consisted of a nail immersed in a bottle partly filled with water. Instead of using a bottle with an outside coating, Musschenbroeck merely wrapped his hand around the bottle. The jar was more or less grounded through the operator's feet. Good grounding of the outside coating to earth was found to be essential when charging the Leyden jar.

9-1
Lidless Leyden jar condenser.
Physikalische Technik, vol. 2, Frick, 1907

A very good discussion of the history of the Leyden jar's development, including the early theories about its action, can be found in *Electricity in the 17th and 18th Centuries* (1979) by J.L. Heilbron.

The most interesting facet of this experiment worth emphasizing to the electrical hobbyist is that the jar was not discovered by following the accepted rules of electricity of the day, but rather by being unaware of them. This fact of history provides good "food for thought," even today. We generally think that worldwide communication

121

assists scientific innovation, yet I believe that incomplete communication can, and often has, allowed an innovator to think and try an experiment that those steeped in orthodoxy refuse to consider. Complete communication does foster a bland uniformity, and it cannot always give impetus to creative developments.

Dielectric constants

The amount of energy, which is stored in a condenser, is expressed in the formula:

$$E = \tfrac{1}{2}\,CV^2$$

C, the electrical capacitance, increases with the area of the jar coatings, with the kind of insulator, or *dielectric*, and also with the thickness of the insulation separating the coatings.

V is the *applied voltage*, or *potential difference*, that is produced between the two coatings.

Each dielectric is assigned a number that varies with the internal structure of the material. That number is termed the *dielectric constant*. Some typical dielectric constants are given in Table 9-1. The higher the number, the greater the capacitance, and therefore, the larger the stored energy.

Table 9-1
Dielectric constants for insulators

Insulator	Dielectric constant
Plate glass	6.8–8.4
Pyrex glass	4.1–6.1
Hard rubber	3.0
Polystyrene	2.5
Shellac film	4.0
Spar varnish	4.8–5.5
Beeswax (purified)	2.9
Paraffin wax	2.5
Wood (dry)	2.0–5.2
Air	1.0

So, a Leyden jar made from a glass bottle will produce a hotter spark than one using a plastic bottle of the same dimensions. Since many types of glass attract moisture at room temperature, you should first test its ability to hold a charge before actually building a condenser.

When choosing jars, flint, crown, and Pyrex glass are recommended. The jars should have openings large enough to admit your hand (for applying the inner coating), and the glass should be clear and bubble-free. To reduce electrical stresses, the jar should be cylindrical and have rounded edges at the bottom.

Once you select the jar, clean and dry it, finish with a lint-free cloth dipped in alcohol to remove grease.

Warm the jar in an oven to about 100°F, remove, and rub the side briskly with a silk or satin cloth. Bring this rubbed area of the glass near the pole of your electroscope to test for the presence of a charge. If the charge is favorable, you should coat the jars, while warm, with varnish inside and out; shellac varnish, polyurethane, or copal varnish is recommended. These finishes prevent leakage from moisture. Only Pyrex glass jars need no coating of varnish because they are not hygroscopic. Cure the jar in a dry, dust-free warm room.

Methods of coating jars

You can imagine how shocked the discoverers of the first Leyden jar were to find that the human hand makes a very painful and too effective coating for an electrical capacitor; this quickly gave them the incentive to look for better conductor coatings. The simplest method used metal powders because these could be applied through the small opening in the jars and easily coat the wall and bottom as a unit. If you choose plastic or glass jars with an opening big enough to slip a hand through, then place masking tape around the inside circumference using Fig. 9-2 to determine the height of the coating. The tape provides a smooth edge for the glue.

Allow 1 to 2 ounces of Elmer's wood glue to thicken in air until it flows like molasses. Using a brush, evenly coat the bottom and sides of the jar up to the tape line. Remove or spread out excess drops of glue, and let the jar stand for about 5 to 8 minutes, until the coating is tacky. Now pour in metal filings or fine powder. Copper, brass, iron, or tin is recommended. Do *not* use lead powder because of the breathing hazard created with this coating. Roll the jar by hand until the inside is

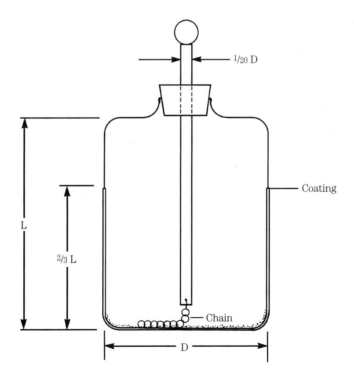

9-2 A well-proportioned Leyden jar.

coated evenly. Dab any open spots with glue and sprinkle them with metal. Immediately remove the tape and let the jar cure for a day. Turn the jar upside down and tap to remove excess metal. Plastic containers are often too smooth for the glue to stick, so they are roughened with 220 grit sandpaper for better adherence. The outside coating may be treated the same way.

By the 1760s, natural philosophers (today called *physicists*) had moved on to thin sheet metal coatings of tin, copper, or lead. Wealthy electrical experimenters, like William Constable of England, had crown glass jars chemically coated with amalgam; this was the same process used for making glass mirrors.

In our time, a good and effective capacitor might employ brass or steel shimstock 0.002 to 0.004 inch thick or even extra heavy-duty aluminum foil.

Varnish thinly and evenly applied to the inside of the jar will secure foil coating, but epoxy glue is better for the stiffer shimstock. Always place factory-cut edges at the top side of the jar to reduce leakage. Keep the shimstock in place with wood splints or rubber bands to prevent separation. The charge leakage can be reduced further by applying melted paraffin to the top edges of both coatings.

Lids, which support the inner metal pole of the jar, were originally made from cork or baked mahogany dipped in a paraffin bath. Figure 9-2 is a general design for a "dry" condenser, which uses metal coatings.

Many jars can be joined together in parallel to form a Leyden "battery," but time is saved by resorting to the "wet" condenser which uses conducting liquids in place of metal coatings. The jars should, of course, be varnished as before. It is important to keep the liquid levels at the same height, both inside and outside, to avoid electrical stress. A description is provided in Fig. 9-3.

Although a layer of linseed oil or castor oil can be used for insulating liquid condensers, I recommend paraffin oil, which has better electrical properties.

Making large capacitors

Warning: Since even a small one-pint Leyden jar gives a very painful and dangerous spark, I recommend that only those with considerable experience build larger ones.

The main difficulty in making large, high-voltage jar capacitors is preventing corona leakage from the top sharp edges of the coatings. Figure 9-4 shows a cutaway construction drawing for a 5-gallon bucket capacitor with lid. Five-gallon buckets are quite uniform in dimensions with stiffener fins molded in near the top. I recommend buying new light-colored ones rather than recycling buckets because capacitors must be quite clean to hold a charge. Hacksaw off the handle, and slip out the two pieces from their supports. Place the bucket on a smooth table, and rigidly support a black fine-point marking pen at the 10-inch level. Slide the bucket against the point, and rotate the bucket to give a horizontal line around the outside; this will mark the top edge of the coating. Roughen all the surfaces to be coated with #220 grit sandpaper, including the bottom inside and outside and up to the horizontal line. Using heavy-duty aluminum foil, cut out discs to fit the inside and outside bottom, spray them with contact adhesive (used for poster board artwork), wait until the adhesive is tacky, and apply. Work the foil discs down with a wood or hard-rubber roller such as used in wallpaper application. Figure 9-5 shows one method of foiling the outer sides. Glue $4 \times 10\frac{1}{2}$

A Liquid Condenser

In the condenser described below, instead of using tinfoil for coating the dielectric, a conducting liquid covered with ½ inch of boiled linseed oil is employed, a practice which obviates all brush discharge. It also has the addi-

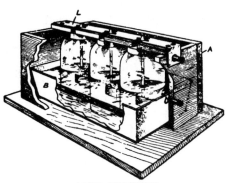

LIQUID CONDENSER

tional advantage of insuring perfect contact between dielectric and coating, thus obviating all uneven distribution of strain and decreasing the liability of a breakdown.

A box (A) of proper dimensions is constructed of paraffined oak. The strip top (L) should have deep notches sawed in it to correspond with the mouths of the jars. A bread pan (B), long enough to hold three jars side by side, is placed in the box and in it three Mason pint jars.

The conducting liquid is then poured to within ¾ inch of the upper edge of bread pan and over it is placed ½ inch of boiled linseed oil. The jars are filled in the same manner. The conducting liquid may be either dilute sulphuric acid (fifteen parts water and one part acid), a strong solution of sal ammoniac, or a salt water solution.

Three connectors are constructed by soldering a short length of No. 8 copper wire to a copper plate. Two supports of No. 12 copper wire are soldered at right angles to the large wire and are so spaced that when the upper one rests on the connecting strip the plate is immersed in the conducting liquid, and when the lower one rests on the connecting strip the plate clears the liquid by ½ inch or more.

The top is then put in place and over it the connecting strip. The three connectors are then slipped in place and the top and connecting strip are screwed to the box. A binding post is mounted on the bent end of the connecting strip. Another binding post is mounted on the same end of the box and fastened to it is a No. 12 wire which makes contact with the bread pan.

The capacity of one of these units of three jars is .00234 microfarads when all the jars are connected, .00156 when two are connected, and .00078 microfarads when one jar is connected, for it is a well known principle that when a number of condensers are connected in parallel the total capacity is equal to the sum of the individual capacities of the several condensers.

9-3 Liquid condensers in parallel. *Popular Electricity, 1912*

inch foil strips with ½ inch extending under the bottom edge, overlapping the foiled bottom to give electrical contact. Proceed around the circumference, overlapping strips vertically for good contact. It may be easier for the inside coating to substitute adhesive foil tape 2 inches wide and apply it in vertical strips with overlap. Be sure each strip contacts the foil disc at the bottom. Remember that plastic buckets become charged as you handle them. Clean with alcohol-dipped lint-free cotton cloth to remove grease, dust, and hair before applying foil coatings. Purchase two macrame rings made from brass-plated steel ³⁄₁₆ × 12 inches in diameter. These are placed at the top edge of the foil, inside and outside, to reduce corona charge leakage from the sharp edges at the

10-inch-high level. Cut each ring, overlap the ends when in place at the top edge of the foil, and mark the excess to cut off. The ring ends should butt together. I next slip a section of ³⁄₁₆ inch i.d. × ⅜ inch long brass tube over the butt joint and silver solder it to give a strong ring. Each ring is tapped into position to give a snug friction fit; each ring should just cover the top edge of the foil. Cement each ring in several spots with super glue, and coat the inside ring with red high-voltage varnish. Figure 9-6 shows one such capacitor and a sample exterior ring. Notice that the outside ring is sheathed with polyethylene tubing ¼ inch i.d. × approximately 37½ inches long. I slit the tubing, which has a natural curl, on the inside curve lengthwise. This slit tube slips over the outside steel ring and snaps into place with a simple butt joint. This method is more effective at preventing leakage from the outside coating. I place a small amount of crumpled foil inside the bottom for the central pole to sit on (see Fig. 9-4). This spreads the current so that hot spots do not melt holes in the foil. PVC or clear butyrate tubing is slipped over the copper tube to reduce leakage from the pole. A simple transparent lid made from acrylic (14 inches square) covers the bucket and keeps it clean inside. The top of the central pole is capped with a terminal such as a 1½- to 2-inch Chinese exercise ball. A capacitance meter gives a value of about 0.0008 microfarads

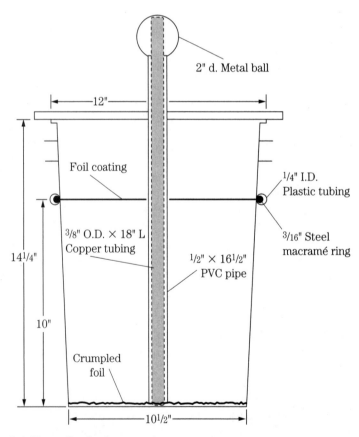

9-4 Five-gallon bucket capacitor.

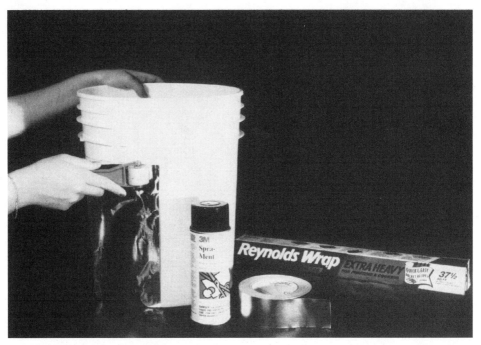

9-5 Applying vertical strips of foil.

9-6 Finished bucket capacitor with corona suppressor.

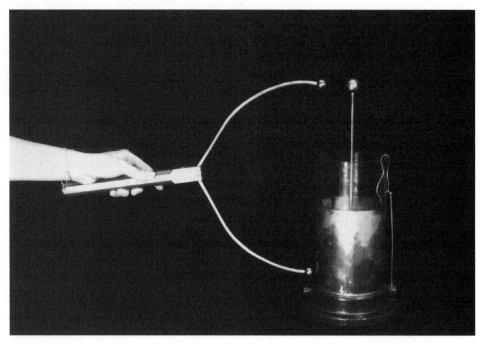

9-7 Proper way to discharge Leyden jars.

for the finished product. Increasing the height will increase the capacitance, but there is a tradeoff because maximum voltage is reduced as the foil coatings approach the top of the bucket. As designed, the distance between coatings over the top is 8 inches (which matches the maximum potential of a 12-inch sectorless Wimshurst).

Be sure to wear ear protectors when discharging capacitors.

If you do experiments with *lidless* Leyden jars, you will notice a particular odor, after repeated discharges, that fills the room. Aluminum, copper, tin, and iron coatings all give off a characteristic smell produced because capacitor discharges vaporize the metal and repel it like a vibrating bell out into the air. This odor I call the "breath" of the capacitor. Figure 9-7 shows the use of discharge tongs. Touch to the outside (grounded) coating first; then approach the central terminal.

Like the electroscope, the condenser is a very important tool for research. Even though it appears to be simple and straightforward in design, a careful investigation of it can provide a wealth of knowledge about stored energy and the nature of electrification in dielectrics.

Design modifications of Leyden jars

One of the least-explored areas of the Leyden jar involves making subtle changes in its materials and geometry. I have already indicated experiments with the electroscope that involve jar design changes. It follows that it is possible to vary the quality of the electricity and the behavior of the jar.

One of the recent tendencies of scientific research is what might be considered a drift toward increasing sterility and uniformity. We picture the scientists in clean white suits working in a windowless laboratory, well isolated and insulated from the natural environment. This tendency appears to be our collective philosophy—almost an unconscious attitude of how we view science as a mental discipline.

In the eighteenth and nineteenth centuries, a different wind was blowing in science. There were no specialists in the front lines of research, but rather amateur investigators, broadly educated and highly skilled, whose main desire was to satisfy their intellectual curiosity. They were not called *physicists*, but rather *natural philosophers*.

We would do well to revive the spirit of inquiry from those earlier days. Do not be afraid to cross over into disciplines outside of, or seemingly separate from, physics.

For example, when exploring creative design concepts involving the Leyden jar condenser, you might break from the idea that the coatings must be perfect metallic conductors in the form of uniformly thin sheets. Consider not only coatings made with metal powders, turnings, or shot, but also intermetallic compounds and semiconducting stone powders. In like manner, consider materials other than ideal insulators, like glass or plastic. For example, try semiconductors that become insulators at elevated temperatures. This area was explored from 1759 to 1762 by Edward Delaval of Cambridge, England. He found that stone, clay, and charred wood, when heated, lose their moisture and become good insulators.

When doing creative work you should get out and explore your natural surroundings for inspiring ideas. You are working with natural energies in this branch of electrical science, and these same energies shape and form the environment. It is obvious that natural conditions are neither sterile nor uniform. Note the shapes of sea shells, insect bodies, seeds, and so on.

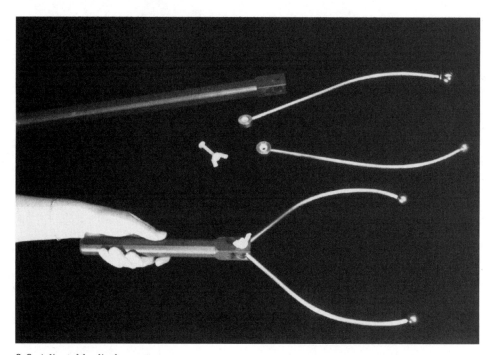

9-8 Adjustable discharge tongs.

The demands of contemporary physics show that entropy can only increase—meaning that the universe is running downhill. When we consider the energetics of biological and weather systems, we see abundant evidence of a run-up potential leading to increasing order. The new science of chaos is a study of the increasing order in natural systems. Both increasing entropy and decreasing entropy are needed to give a full picture of the equilibrium-seeking natural energies in the environment.

Finally, I give a word of warning. A 1-pint Leyden jar, fully charged, packs a wallop sufficient to knock you off your feet. Show a charged large jar (1 to 5 gallons) the same respect you would give to any powerful explosive. Use the special discharging tongs for safety's sake. Never use wires or screwdrivers (see Figs. 9-7 and 9-8). Finally, store the jars with jumper wires joining the inside and outside coatings. This method ensures that residual charges can't build up, as often happens when using glass as a dielectric.

10
CHAPTER

The electrophorus

The last basic instrument discussed in this book is essentially a modified condenser with dissectable parts. When the coatings and dielectric are manipulated in a special way, the condenser becomes a charge-dispensing device, called an *electrophorus* (Fig. 10-1).

10-1 The basic electrophorus. *Physikalische Technik*, vol. 2, Frick, 1907

Johannes Wilcke, who worked with dissectable condensers, developed this device in 1762. The device was later called the *perpetual electrophorus* because once the dielectric was charged, a seemingly endless quantity of electric charge was produced by special manipulation. In 1775, Alessandro Volta named the device and perfected its manipulation. His theory of its action was not as clear as Wilcke's explanation, however (Fig. 10-2).

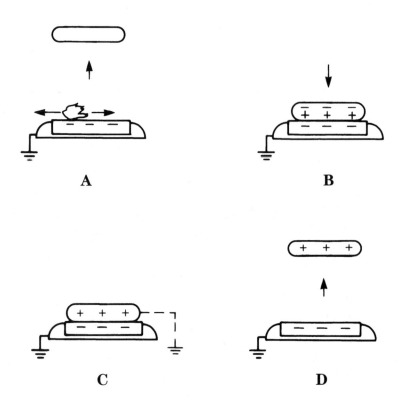

10-2 Steps in charging the electrophorus.

In Fig. 10-2:
A. The dielectric cake is charged by rubbing it briskly with a cloth. The cake is resting on a metal base that remains grounded.
B. The metallic cover is placed on the cake.
C. The like-induced charge is conducted to the ground. Then the connection is broken.
D. The cover with the opposite charge is lifted from the cake.

Because the cake is an insulator, it can't give all its charge to the cover or to the base by conduction, so its electrification, once produced, can persist for a long time. To operate the electrophorus, discharge the cover, replace it on the cake, temporarily ground it, and then remove it. Operated properly, a continual charge supply is available, hence the name *perpetual electrophorus*.

Presently, the action is said to be caused by electric induction across surfaces in close proximity. Extra mechanical work is required to remove the cover because of its electrical adherence to the cake. Not well understood is just how the resin cake, an insulator, can continue to supply charges over a long period of time.

Notice that the electrophorus is the electrical analog to a permanent magnet. The "keeper," placed on the magnet, preserves its power just as the cover plate preserves the cake's charge. Cakes have been known to preserve an initial charge for up to a year!

The grounded electrophorus

The first simple modification to the electrophorus was mentioned by J. Phillips in the *Philosophical Magazine* in 1833. Until that time, the experimenter's finger was used to temporarily ground the cover (see Fig. 10-2, step C). Phillips had the bright idea of pasting a thin, narrow strip of tin foil across the entire face of the cake, each end in contact with the metal sole plate. The foil automatically took the place of finger grounding, but the funny thing is that the foil short-circuited the plates of his dissectable condenser! Even so, the resin cake did not discharge.

Another method Phillips mentioned was to bore a small hole through the cake's center and place a metal plug in the hole, contacting the sole flush with the cake's top surface. Either method must allow intimate contact between the cover and the cake.

Electrical shock from a sheet of paper

One modified form of the electrophorus uses thin film semiconducting surfaces in place of thick resin cakes.

Get a metal japanned (i.e., with an enamel coating) tea tray with rounded edges and a flat bottom. Then support it on two cakes of paraffin or dry glass tumblers. Cut a piece of heavy brown wrapping paper so that it covers the tray's bottom. Near the fireplace, hold the paper until it is quite warm, but not scorched. Hold one end down on a dry wood table while giving it a dozen strokes with a latex rubber glove or neoprene balloon skin. Now, bring the sheet by two corners over the tray. It will fall like a stone. If you touch the tray, a noticeable shock will be felt. Note that this experiment must be done in a warm, dry room for best results.

What is interesting here is that by heating paper above room temperature, you can measurably increase its resistance by driving moisture from its pores.

Experiments with the thin film electrophorus were extensively covered by Professor Joseph Weber of Ingolstadt, Germany. His main work on this subject, *Abhandlung von dem Luftelektrophor* (1779), contains a wealth of information. Unfortunately, it is written in the old German script, which makes translation difficult. His simple and inexpensive designs employed fabrics, such as linen or silk, tightly stretched on wooden frames. These frames were placed in the vertical plane near the fireplace. Once warm and dry, considerable amounts of electricity could be drawn off after the membranes were briskly rubbed. It is not clear if the stretched fabrics were first varnished (after the method of "oiling silk)."

Making the traditional cake electrophorus

The original electrophorus designs used dry glass handles for holding the cover plates, which must have well-rounded edges. To size the cover plate, make its diameter about $9/10$ diameter of the resin cake.

When the electrician could not afford the expense of having a metal cover made, a fine-grained hardwood, such as mahogany, was shaped and planed smooth, and metal foil was pasted on it. The resin cake (dielectric) must be dense, smooth, and hard, and never cast less than ½ inch thick. Usually the thickness is 5 percent of the diameter of the cake.

Both the cover and the cake's mating surfaces must be perfectly flat; when pressed together, a slight vacuum forms.

As far as electrical output is concerned, a resin cake 8 inches in diameter can deliver about 1 microcoulomb of charge to its cover at each contact. The spark length is normally one-tenth the diameter of the cake. For example, an electrophorus cake 24 inches in diameter will produce a 2½-inch spark. Here again, the device should be used in a warm, dry room for best results.

Baking a cake

Although brimstone (as molten sulfur) makes the best cake, it often cracks from temperature changes. Repeated liquefaction before casting and slow annealing after casting offset the cake's brittleness to some extent.

An electrician, William Snow Harris, provided several formulas for resin cakes, one of which follows:

Refined shellac flakes	1 part (by weight)
Venice turpentine	1 part
Resin	1 part

Melt shellac and resin slowly in a covered iron crock placed on a sand heat. When melting commences, add the turpentine and continue stirring with a glass rod. Once the solution is perfectly fluid, add a little coloring, such as ivory black, to produce a nice appearance. Retain the melt in this state until the air bubbles are expelled. An alternative, simpler method replaces Harris' cake formula with sealing wax in stick form. The wax sticks are melted in a mixing bowl over low heat.

Caution: Never melt this combination over an open flame!

A clean casting surface is usually polished marble or plate glass. Place a ring of sheet metal or wood of the desired thickness on the casting plate and place lead weights on the ring's edge to prevent the mixture from seeping out. Pour the melt into the mold, until the ring is filled and level. Allow this mixture to remain undisturbed until the cake is quite cold—overnight is best. If the wood ring must be removed from the cake, use a split ring, bound with a string and lined with foil, inside.

I prefer wooden embroidery hoops for the casting form; wooden basket hoops are also serviceable. Harris did not specify the type of resin used, but either ester gum or rosin would be a good choice. The melt temperature is normally 250°F, and this fluid condition should be maintained for at least 30 to 60 minutes.

When the cake is cold, the surface next to the casting plate, which is perfectly flat, becomes the top surface next to the cover plate. The cake can be placed in a metal cake pan, which serves as the bottom or sole plate. The pan should be electrically well grounded.

Furs (and wool) were originally used to excite the cake surface; but a sheet of silk, satin, latex rubber, ripstop nylon, or flannel will work. All materials must be warm and dry. Using oblique glancing blows sometimes works better than simply rubbing the surface. Now, holding the cover's handle, place it down on the cake, ground it temporarily, as described before, and lift the cover.

Bring your finger near the cover and you will promptly receive a sharp, snappy jolt—a miniature lightning bolt! Use Phillip's method for automatic grounding and you can draw a spark each time you lift the cover (without charging the plate). You can continue this process until you are tired. Some folks find a shock to the system quite exhilarating; not me. Always store your electrophorus with its cover on, to protect it from dust.

The smallest cakes have been only 1 inch in diameter (for charging electroscopes). One of the largest electrophorus cakes ever recorded was made in 1777 by Professor Lichtenberg. This monster was 6 Paris feet in diameter, and the cover had to be applied with a rope and pulley hung overhead. The cake consisted of resin, turpentine, and Burgundy pitch. The spark was said to be 15 inches in length—thick and hot.

Today, the study of the electrification of dielectric bodies centers around *electrets*. The main difference between electrets and resin cakes is that electrets are charged by applying a voltage while they are hardening. No mechanical rubbing of the surface is then required. A typical formula for an electret consists of 45% carnauba wax, 45% rosin, and 10% beeswax. These substances are melted together, as before mentioned.

The author's electrophorus designs

Because of the fire hazards associated with melted wax and turpentine, I've developed simpler methods for making the electrophorus that do not require mixing chemicals. Some modern plastics can take the place of sulfur or wax plates because of their good dielectric properties; Pyrex glass (also called *borosilicate glass*) has such a high silica content that, like quartz, it cannot absorb moisture. In general, the dielectric material chosen for the electrophorus should have a high dielectric constant and a smooth, flat surface with low moisture absorption. A crude test for moisture content may be made by placing a glass or plastic plate sample in the microwave oven for about 30 seconds to see if it warms up. **Warning:** Be sure to use only plastics that are "microwave safe," because some types melt and catch fire. At the restaurant or kitchen supply store decide which metal pans would make a good cover plate. The cover plate should have no sharp bends, have a large rolled or beaded edge, and be measurably flat. Use a steel rule on edge to determine the flatness of the bottom surface. A compact and readily made electrophorus that uses a "deep dish" aluminum pizza pan is shown in Fig. 10-3. The large rolled edge prevents charge leakage. This design incorporates Phillips' automatic grounding method so that step C in Fig. 10-2 is done with grounding screws rather than your finger. The dielectric is a ¼-inch-thick disc cut from mechanical-grade Teflon plate. This disc is fastened to a wood base with a disc of aluminum foil sandwiched between to act as a bottom plate. When assembled, the electrophorus becomes a dissectable plate capacitor. All three grounding screws are joined by the aluminum foil, with one being grounded. The screwheads must be *flush* with the Teflon's surface so that the pan makes flat contact. The pickup handle may be made from delrin, PVC, acrylic, or polypropylene. From the triboelectric series at the end of Chap. 7, select a material that can be used with Teflon to give a good charge. Ripstop nylon fabric cut to a 6-inch square works well. With the base grounded, rub the Teflon briskly to give it a charge, and holding the handle, press the pizza pan down firmly on the Teflon. You may hear

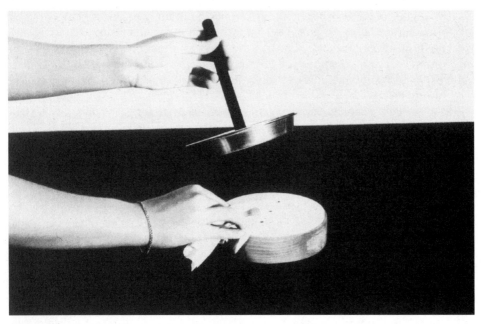

10-3 A 6-inch electrophorus.

a slight click produced by the spark grounding the pan through the screwhead. Lift the handle *straight* up, and you should get a sobering spark to your knuckle ½ to ¾ inch long. Remember that plate and pan must be kept free of grease and dust for best results. Figure 10-4 gives the construction details. Be careful to attach the handle to the pizza pan so that the flathead screw is flush with the surface, and verify this with the edge of a steel rule.

An electrophorus for physics laboratory classes

University physics courses on electrostatics can provide more extensive experiments in the laboratory by using a charge dispenser that gives either a positive or a negative charge. Figure 10-5 shows a compound electrophorus that can be clamped down on the laboratory bench. From the triboelectric series (see Chap. 7) I checked for a material at the opposite end of the list from Teflon. Pyrex (borosilicate) glass plate was found to consistently give a charge opposite to that of Teflon. A 6-inch-square plate ¼ inch thick may be used and is less expensive than a Pyrex glass disc. Figure 10-6 details the construction of this unit. McMaster-Carr Supply Company of Chicago sells both the glass and Teflon plates needed for this project. Their part number for the borosilicate glass is #8476k16, and the catalog number for a 12-inch-square piece of PTFE (Teflon) sheet, mechanical grade, is #8801k511.

The aluminum base plate is formed from approximately 18-gauge soft aluminum and bent into the channel shape with an aluminum angle bolted to one side for clamping to the table. The square glass plate is fastened down with eight mirror mounting brackets called *offset metal mirror clips* with ¼-inch channel. These are sold at

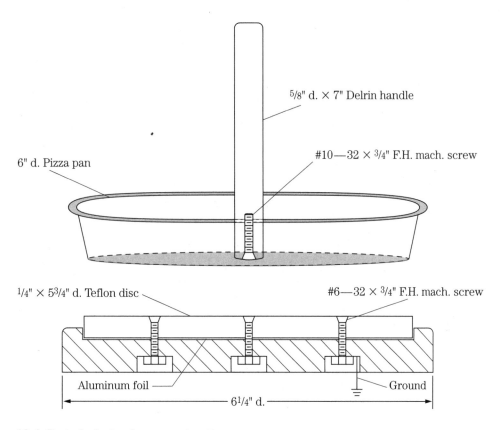

5/8" d. × 7" Delrin handle

#10—32 × 3/4" F.H. mach. screw

6" d. Pizza pan

1/4" × 5³/4" d. Teflon disc

#6—32 × 3/4" F.H. mach. screw

Aluminum foil

Ground

6¹/4" d.

10-4 Six-inch electrophorus construction.

hardware stores and made by the Servalite Company; their part number is #Z44. In the enlarged side view of the same figure, note that a self-adhesive bumper ⁵/₁₆-inch in diameter is placed between each bracket and the glass surface as a cushion and to prevent the glass from sliding. These bumpers are made by Brainerd Manufacturing Company, East Rochester, New York, and are available through the hardware store. The Teflon disc may be fastened down with from one to three #8 flathead machine screws and nuts, care being taken to keep the Teflon very flat. This electrophorus uses the same 6-inch pizza pan and handle described earlier. When charging with the glass plate, slide the pan to one side so that it touches a bracket for the temporary grounding step. Then lift straight up.

Consulting the triboelectric series (see Chap. 7), we see that good results can be expected by rubbing glass with sheet rubber. Surgical rubber called *dental dam* or balloon skin works well. Sheet rubber may be cleaned with mild soap and water, not alcohol. Each dielectric requires a specific material for rubbing to produce a good charge. Experimenting will help you find what works best and gives consistent polarity. When moving from one plate to another, touch the aluminum pan to the aluminum base to remove any remaining charges.

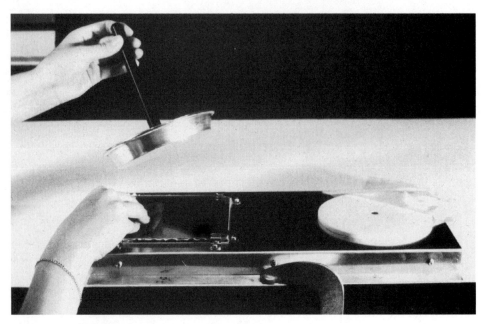

10-5 Dual-polarity electrophorus for the laboratory.

Enlarged side view

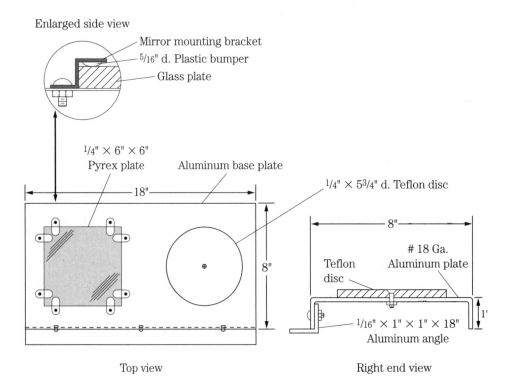

Mirror mounting bracket
$5/16$" d. Plastic bumper
Glass plate

$1/4$" × 6" × 6"
Pyrex plate   Aluminum base plate

$1/4$" × $5^3/4$" d. Teflon disc

—18"—

8"

8"

Top view

18 Ga.
Aluminum plate

Teflon
disc

$1/16$" × 1" × 1" × 18"
Aluminum angle

1'

Right end view

10-6 Details of dual-polarity laboratory electrophorus.

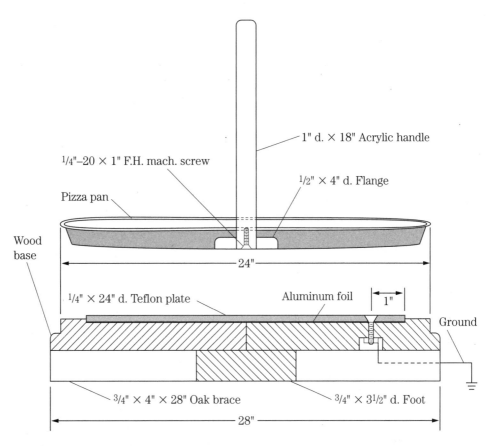

1" d. × 18" Acrylic handle

¼"–20 × 1" F.H. mach. screw

½" × 4" d. Flange

Pizza pan

Wood base

24"

¼" × 24" d. Teflon plate

Aluminum foil

1"

Ground

¾" × 4" × 28" Oak brace

¾" × 3½" d. Foot

28"

10-7 Twenty-four-inch lecture room electrophorus.

A large lecture room electrophorus

A 24-inch electrophorus suitable for the university physics lecture room is illustrated in Fig. 10-7. This is simply a scaled-up version of the 6-inch electrophorus (see Fig. 10-4). The base is constructed using two 1 × 14 × 28 inch boards glued together with a ¾ × 3½ × 28 inch oak brace running perpendicular to the glued joint. Care must be taken to get a flat top surface, which should be tested with a steel rule. This flat surface establishes a uniform base for the Teflon plate so that it closely mates with the pizza pan cover. Two wood discs 3½ inches in diameter serve as outrigger feet to stabilize the wood base. A 24-inch disc of aluminum foil is placed; then a 24-inch disc of mechanical-grade Teflon is placed over the foil and secured with eight #10-32 × ¾ inch flathead machine screws and nuts around the perimeter and one screw in the center. The perimeter screws are placed 1 inch in from the edge of the Teflon as shown, and one is grounded through a small wire.

In choosing the metal cover plate, as a rule, large aluminum pizza pans are not flat, so, if possible, use a 24-inch steel or stainless steel pizza pan with a large beaded edge. The handle, made from acrylic, PVC, or polypropylene, needs to be 18 inches so that your hand and body are well away from the pan. A 4-inch-diameter plastic flange reduces side stress where the pan joins the handle. The technique for using this device is to connect the base screw to ground and touch one finger to one of the screwheads while vigorously rubbing the Teflon with a sheet of ripstop nylon. (By grounding yourself, you will not get shocked while rubbing the surface.) Holding the handle near the top, place the pan down. You should hear a click as a small spark passes from pan to base screw; lift *straight* up and bring the pan near a clean, flat, and vertical grounded metal surface, such as a metal cabinet. **Caution:** Avoid using near any sensitive electronic equipment, including motor vehicles. Properly done, snappy sparks about 3 inches long and diffuse, quiet sparks up to 8 inches long are produced. For the materials involved, spark length will be about ⅛ the diameter of the dieletric! Remember that the Teflon only needs rubbing *once,* not every time the pan is placed. The dielectric can store this charge for months if it is kept covered with the pan.

A simple charge indicator placed on the cover shows no charge when the cover is down and a high charge when the cover is raised. Figures 10-8 and 10-9 show "Judy," first at ground potential and then at 50,000 volts.

From the toy store, "Judy" came with blond plastic hair that I replaced with ⅛-inch strips of facial tissue. A thin strip of adhesive copper foil runs from her head to her feet to transfer the charge. The strands of paper are attached with wood glue. After her feet are glued to a metal base, "Judy" is transformed into a very pretty electroscope!

This experiment vividly illustrates the relationships between voltage, charge, and capacitance. Remember that the electrophorus is basically a dissectable flat-plate capacitor with a metal cover, an insulator plate, and a grounded metal sole plate. Since

$$\text{Voltage } V = \frac{\text{charge } Q}{\text{capacitance } C}$$

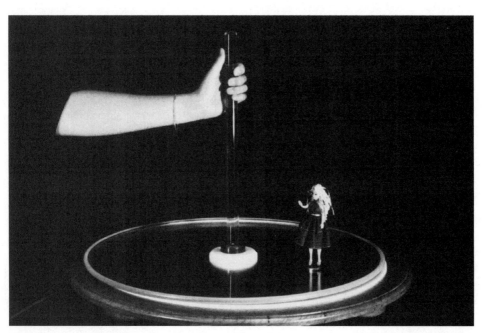

10-8 Folks, meet "Judy."

10-9 "Judy" all charged up but no place to go.

for a given charge on the plates of the capacitor, as the cover plate is raised, the capacitance must decrease. This decrease in capacitance appears as an increase in voltage or potential difference between the top and bottom plates. The charge on the cover plate becomes unbounded, or *free,* and appears as a spark.

Figure 10-10 shows two fanciful but entertaining charge indicators. "Eggbert" also uses thin strips of facial tissue glued to a ping-pong ball having two big dancing eyes. The "octopus" (or "spider") is suspended from a nylon or silk thread so that he shows his potential well. Both "Judy" and "Eggbert" say they really get a charge out of experimenting with the electrophorus.

One instrument called a *proof-plane* is useful for sampling a small amount of charge from a surface. An insulated thick and well-polished metal disc should be used for charging electroscopes, whereas touching the scope directly with the cover pan will damage the scope's sensitive leaves. Figure 10-11 shows two sizes of planes. When used, the proof-plane's charge polarity should be the same as the surface touched; hence it "proves" the polarity of the surface sampled. In Figure 10-12, the metal disc must have well-rounded edges to hold a charge; it can be joined to the handle using super glue; I recommend a section of heat-shrink tubing to strengthen the thin plastic rod and thereby prevent breakage.

Semiconductive stones

There is, however, a tiny "fly in the ointment," which appeared in a press release in *Popular Mechanics* in 1935. What goes on here? After I studied the article and its accompanying photo for some time, I guessed that a peculiar variation on the

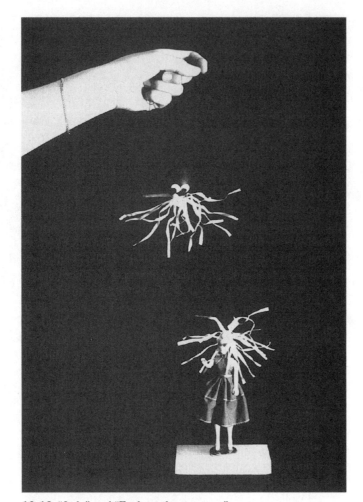

10-10 "Judy" and "Eggbert the octopus."

electrophorus was invented back then. The photograph showed the inventor with a wood and rubber "mushroom" that was claimed to produce an electrical charge. According to the article, a spiral of metal was in or on the electrically charged circular glass plate. The spiral's outer end could be grounded through the handle (Fig. 10-13).

A 12-inch-diameter wood "mushroom" device is covered over with a sheet of rubber. This convex shape and the membrane would make a very good means for exciting the plate of an electrophorus. The thick white block, which appears to be marble, looks like a sandwich of three pieces. Could there be a metal spiral inside it also? The article stated that when the block is "rubbed with the mushroom for five minutes, sufficient electricity is generated to keep a neon tube glowing for eight to twelve hours."

The article also stated that when a neon gas tube was moved back and forth in front of the plate, the tube would glow brightly.

A really strange question is this: If a neon light can be energized for several hours without rubbing the plate, should we reconsider the theory of electrification,

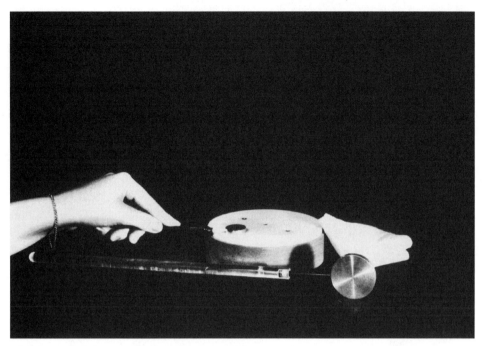

10-11 The proof-plane samples a charged disc.

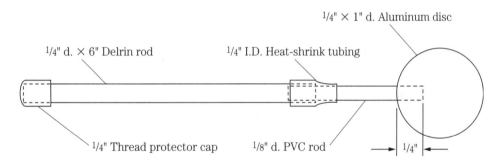

10-12 Proof-plane construction.

which relies on static charges? Does the electrophorus have a persistent dynamic energy flow that requires an initial rubbing action to get it started?

The electrical properties of semiconducting stones remain largely unexplored. However, accurate studies of these stones might be necessary in order to explain the operational theory of the electrophorus.

Alessandro Volta wrote a report in the *Philosophical Transactions* (1782) on his work with semiconducting materials. One discovery was that a conducting plate retains a charge better when it is in contact with a poorly insulated body (semiconductor) than when resting on a resin cake (which is well insulated)! This discovery became known as *Volta's Paradox*. One of the semiconducting materials best suited

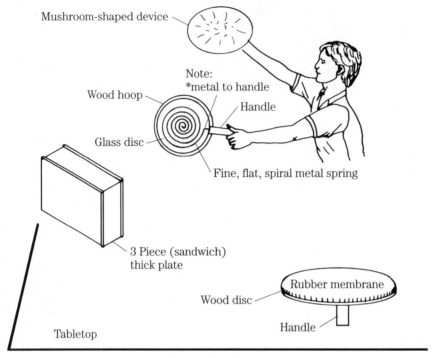

Mushroom-shaped device

Note:
*metal to handle

Wood hoop

Handle

Glass disc

Fine, flat, spiral metal spring

3 Piece (sandwich)
thick plate

Rubber membrane

Wood disc

Handle

Tabletop

10-13 A charge-radiating electrophorus. *Popular Mechanics, 1935*

for this experiment was a polished plate of white Carrara marble, warmed up to drive off moisture.

Related to Volta's Paradox is the *Johnsen-Rahbek Effect*, described in 1923. This discovery concerns the tendency for certain kinds of stone to adhere strongly to metal plates when a high-voltage direct current is applied. The adhesive force, from electrostatic attraction, was found to increase with the fifth power of the applied voltage. Several patents, such as 1,533,757, April 1925, were issued to Johnsen and Rahbek. They preferred stones that were finely porous and hygroscopic—slate, marble, steatite, agate, limestone, flint, and jasper.

The physics of electrical adhesion is not well understood. As far as the mushroom generator from Berlin is concerned, no U.S. patent was found, but there is a possibility that a German patent was issued. If so, it probably would have appeared between 1930 and 1939.

Some avenues for electrical pioneers

Although it is one of the simplest electrical generators, the electrophorus should be considered as important as the electroscope and the condenser to basic research on the nature of electrification. The long-persistent charged condition of the dielectric is still not well understood. Because of this general lack of understanding, I have explained in detail the traditional methods of making this generator.

Several questions present themselves at this point:

- How are the plate's conditions (smoothness and thinness) important to electrification and its persistence?
- In semiconducting plates, how do surface résistance, bulk resistance, porosity, moisture content, and temperature affect electrification?
- What are the best materials and what is the best shape for the rubbing device used for charging the electrophorus's plate.
- Should the sole or bottom plate of the device be a solid sheet of metal or should it spiral from the center to the circumference? If it is to spiral, would an arithmetic or a logarithmic shape be best?
- How does the length of time spent rubbing the electrophorus affect the persistence of electrification?

When returning to nature for examples, note how often multiple layers are used in constructions. The layers in onions, tree trunks, rocks, minerals, and soils suggest experiments with layered electrophorus plates using alternating films of dielectric and semiconducting materials, for instance glass and wax. Experiments along these lines can help us to visualize what takes place during the condition we call *electrification*. Until this is visualized, we cannot really understand, and therefore cannot design to enhance this process.

A more complicated generator, such as the Wimshurst machine, will use the same basic principles found in the electrophorus. However, the simple must be understood before the complex can be explained.

Summary

The electrophorus, when it first appeared, caused almost as much consternation as its predecessor, the Leyden jar. As a result, some major theories were made or dashed.

In visualizing electrification, we might, for example, picture the plate of an electrophorus as covered with depressions. Each depression is occupied by a freely suspended marble. A glancing impact to the surface will set several of the marbles spinning like gyroscopes. This persistence of motion might serve as a dynamic model of what happens during electrification at the molecular level.

Although the costs of making an electroscope, Leyden jar, and electrophorus are quite reasonable, you will need to master some new skills in this work. I advise you to keep your notes in a journal, especially recording the daily weather conditions. The amount of knowledge gained with these instruments is limited only by your ability to create original experiments. As Ernest Rutherford said years ago, "An ounce of thought is worth a ton of equipment."

11
CHAPTER

Electrostatic motors

Many subjects in high-voltage work are great science fair and college graduate research projects. Because of the subtle variations possible in the design and materials used in these experiments, a single avenue of information can provide a wealth of knowledge for the researcher over many years.

An excellent book that discusses the history, types, and principles of operation is *Electrostatic Motors* by Oleg Jefimenko (1973). This electrostatic motor book traces electrostatic motors from their beginnings in the eighteenth century up to about 1970. An excellent bibliography is included.

Conventional electric motors use electromagnetic force to convert electricity into torque. However, a less-explored class of motors produces torque by transferring energy electrostatically by way of contact, corona discharge, and/or induction. Because it is difficult to accumulate large electrostatic charges without breaking down, electrostatic motors have proven to be most effective where small size and high speeds are needed. Electrostatic motors can operate on currents as low as one-billionth of an ampere, and have even been operated directly from atmospheric electricity when an antenna is used.

Because of the extreme simplicity of construction, a quiet dependability of operation can also be expected. Figures 11-1 and 11-2 show examples developed in the 1890s.

One prolific experimenter with electrostatic motors at the turn of the century was Howard B. Dailey. A sample of his work is described in Fig. 11-3. Another type, working from atmospheric electricity, is shown in Fig. 11-4. In this case, the antenna needs to be well insulated from moisture because of the small charges involved. In the event of thunderstorms, this same antenna must be grounded well for safety's sake.

Building an electrostatic motor

In my experiments with electrostatic motors, I used a setup like Dailey's, except that my rotors were constructed with thick fiberglass or acrylic discs 2 to 4 inches in diameter.

11-1
Disc-type electrostatic motor.
Scientific American, 1891

11-2
Cylindrical electrostatic motor.
Scientific American, 1891

Coat the rim of the rotor with high-voltage (corona) varnish. When the varnish becomes tacky, roll the disc edge in a pile of #60 grit aluminum-oxide grains (sand blaster's abrasive). These grains make the rotor edges semiconducting, and give a large surface area for charge storage. Instead of thick discs, drums of acrylic can be used. The larger surface area means greater torque for a given diameter.

The following experiment was first described in my article for *Electric Space-craft Journal*, January–March 1992.

The Jumping Electrostatic Top
by R.A. Ford

In trying to achieve the jumping electrostatic top, Mr. Ford details his experiments using a homemade sectorless Wimshurst generator of his design. He and his research associate, John Richter, discovered that small discs supported by a single point were the most effective. The question of why the top suddenly jumps up is left unresolved.

Static electric top

The continuous action electric top here described furnishes an interesting demonstration of static electric attraction and repulsion.

The top is a disk of stiff mica $4^5/8$ inches in diameter mounted between two small buttons of wood or vulcanite upon a slender axis made from a piece of steel knitting needle. The pointed lower end projects $1/8$ of an inch below the disk and rests in a small indentation worked with the point of a file in the upper end of the vertical glass standard made from a piece of druggist's acid rod. To give the disk a finished appearance, a broad band of red insulating varnish made by dissolving red sealing wax in alcohol is applied to the disk's circumference.

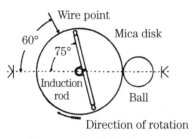

Diagram of Static Top

At the edge of the disk and as close to it as possible without touching is a $1^1/2$ inch polished brass or aluminum ball carried by a second support of glass. Rising from the base of the instrument at the left is a curved brass discharge wire arranged so that its pointed upper end approaches the disk's edge very closely from a horizontal distance, at a place about one third the circumference of the disk from the ball.

Beneath the disk, attached by a small brass fixture to the glass stem near its top and on the side nearest the ball, is a polished metallic induction bar with rounded ends, made of $1/4$-inch round brass rod. This rod is $4^5/8$ inches long and is so placed that its upper surface is about $5/8$ of an inch below the disk. The angular position of the induction rod and wire discharge point relative to the ball is important. The most effective arrangement is that shown in the diagram, in which the wire point and induction rod make angles respectively of 60 and 75 degrees with an imaginary line drawn through the centers of disk and ball.

Static Electric Top

In operating, the insulated ball is connected to the negative pole of a static machine—that pole which gives the brush effect on its collecting combs—the positive pole being joined to the discharge wire through the binding post on the base. The disk, held in position for a few seconds with the fingers placed lightly upon the upper end of its axis, immediately begins a swift rotation, when the finger may be removed and the top will spin at high speed. That position of the disk at the discharge wire receives along its edges and adjacent surface position electrification, causing repulsion from the point with simultaneous attraction from the negatively excited ball. Rotation ensues, the charged sections of the mica arriving at the ball, giving to it positive electricity and receiving negative in a hissing stream of minute sparks. These parts are impelled forward by the similarly charged ball, until within the attracting influence of the positive wire when the cycle is repeated.

The precise nature of the influence exerted by the brass rod below the disk is somewhat obscure. Through some inductive process its presence seems to effect a certain needed balancing of the acting forces; without it the top indulges in violent gyrations and soon tumbles off.—H.B. DAILEY.

11-3 Continued.

Static Motor

Here is a contrivance which will amuse and interest the young electrician. It consists of a small piece of number fourteen gauge copper wire (A), balanced on a pivot (B). Two wires (C) and (D) are supported so that their ends are very close to the ends of wire (A). Connect the lead-in from an aerial to one of these wires, the ground wire to the other, and if everything has been properly made, and if there is a sufficient charge of static on the aerial, the movable wire (A), or armature, will commence to revolve.

For best results (A) should be about 1½ inches long, the distances between the ends of the stationary wires and the ends of (A) being about one thirty-second of an inch. Contributed by S.C.W.

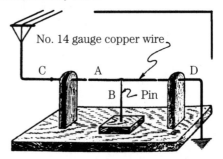

A static motor operated from the ether waves collected by an aerial.

11-4 Atmospheric electric motor. *Practical Electrics*, 1924

Introduction

I first became acquainted with this experiment in the book *Ether-Technology* by Rho Sigma early in 1983.[1] My research associate John Richter and I were working with our homemade sectorless Wimshurst generator, which had 24-inch acrylic discs. In achieving the jumping electrostatic top, we tried many different electrostatic motors, and settled on small discs supported on a single point as being most effective.

Jumping top background

The original French article to which *Ether-Technology* refers appeared in *Revue Francaise D'Astronautique #3*.[2] Therein Dr. Marcel J.J. Pages discusses a disc

---

1 Rho Sigma. *Ether-Technology: A Rational Approach to Gravity Control.* Lakemont, Georgia: CSA Printing & Bindery, 1977. New printing: Clayton, Georgia: Tri-State Press, 1989, pp. 84–5.

2 Pages, Marcel, J.J., "Gravitation Antiponderale," *Revue Francaise D' Astronautique #3*, 1967, Figure 8, p. 9.

electrostatic motor shown in Fig. 11-5. It was invented by H.D. Ruhmkorff (famous for his "Ruhmkorff coils"), and built by his associate, the French scientific instrument maker and inventor, Eugene Ducretet.

This invention was first illustrated in slightly different form as shown in Fig.11-6 by M.E. Mascart in 1876.[3,4] Returning to Fig. 11-5, it consists of a single, thick mica disc, approximately 6 inches in diameter, which spins on a fine sharp point. One of the pointed electrodes below the disc connects to the earth, and the opposite electrode connects with the high-voltage terminal of an electrostatic generator.

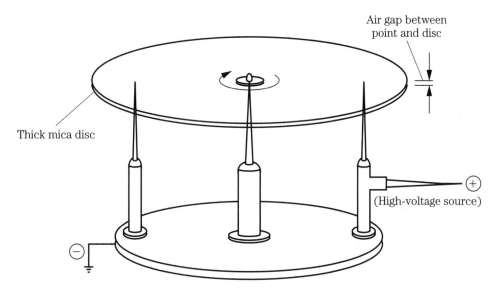

11-5 Ruhmkorff's electrostatic motor.

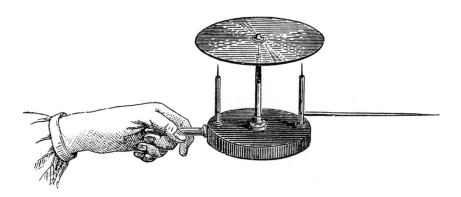

11-6 Ruhmkorff's hand-held corona motor. *Traite' d' Electricite' Statique*, vol. 1, M.E. Mascart, Paris, 1876

3 Mascart, M.E. *Traite D' Electricite Statique*. vol. 1, Paris, France, 1876, Figure 69, p. 179.

4 Jefimenko, Oleg D. *Electrostatic Motors: Their History, Types and Principles of Operation*, 1973. Electret Scientific Company, Star City, West Virginia.

The disc develops a high speed of rotation and at a certain point instantly jumps up a fraction of an inch and off its support. The oddity is that the electrostatic forces are directed only in the horizontal plane, so whence the vertical component force? Knowledge of the jumping disc experiment may date back to the latter eighteenth century. The names of the natural philosophers Ben Franklin, Michael Faraday, Wilfrid de Fonvielle, and later Abbe Laborde have been mentioned in this regard.

1983 experiments

Although my associate, John Richter, and I tried Ruhmkorff's design using two pointed electrodes, we concluded with the geometry described by Howard B. Dailey in *Popular Electricity Magazine*, 1912[5] as shown in Figs. 11-7, 11-9, and 11-10.

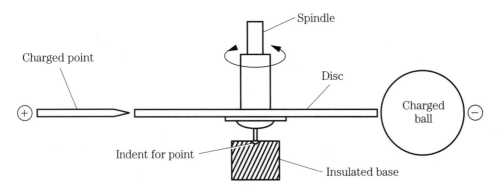

11-7 Howard B. Dailey's static electric top.

The asymmetry of the electrodes greatly improves the performance in both speed of rotation and torque. Dr. Oleg D. Jefimenko's book, *Electrostatic Motors: Their History, Types and Principles of Operation*, is an excellent introduction to electrostatic motors.[6] While there is a good deal of qualitative information on the subject, little has been done on the quantitative side until recently. Even theories of the forces running electrostatic motors that can optimize design are scarce or incomplete. Corona motors may run because of the momentum of ions streaming from points to rotor, but this is disputed by Dr. Stefan Marinov, whose calculations lead him to conclude that coulomb attraction and repulsion are the main forces causing torque.[7] (Observations of the wheel acceleration appear to be stronger than what an ion "wind" would create.)

We first cut mica discs from "isinglass" windows (used in wood stoves). This was about 0.020 inches thick. We stacked several discs together to gain in thickness. These did not spin well and usually stuck in one spot. We then tried white fiberglass washers, 3/32 inch thick × 3 inches in diameter, supported as shown in Fig. 11-8. The plain disc worked well and spun at good speed.

5 *Popular Electricity Magazine*, Chicago, Illinois, 1912.

6 Jefimenko, *Electrostatic Motors*

7 Marinov, Stefan. *The Thorny Way of Truth, Part V*. Graz, Austria: East-West Publishers, 1989, pp. 25–6.

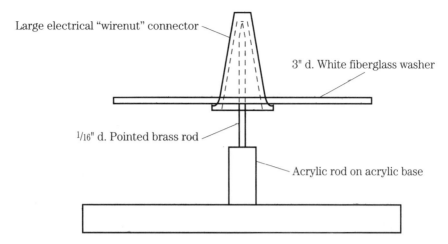

Large electrical "wirenut" connector

3" d. White fiberglass washer

1/16" d. Pointed brass rod

Acrylic rod on acrylic base

11-8 Fiberglass disc trials—1983.

Next we painted spokes onto the disc using a soft graphite pencil dipped in shellac. These worked better. We then rolled the edge of the disc in red, high-voltage insulating varnish and then in various semiconducting powders, getting our best results with medium grit aluminum oxide abrasive grains (sandblaster's abrasive). An acrylic disc with the edge coated in such grit is shown in Fig. 11-9.

11-9
Acrylic disc with annular groove coated with aluminum oxide grains.

In gradual steps we finally concluded that semiconducting discs gave better results than strict insulators. Fiberglass washers treated in this way spin at roughly 5000 rpm if well balanced.

Finally, we tried a miniature aluminum oxide grinding wheel (¼ inch thick × 2 inches in diameter), which was much more massive than the fiberglass disc. These are sold at hardware stores as accessories for electric drills. They include a steel arbor (spindle), which we modify to make into a top supported on a single point. These tops are started with a quick twist of the fingers, then the generator is started. The precession of the spindle decreases as speed increases; we estimate a rotor speed of 5000 rpm. At certain high speeds, sparks pass around the circumference of the wheel or across the wheel between electrodes. About three times John and I saw the top jump up and off its support. We cannot explain this phenomenon.

1992 experiments

In May 1992, interest was expressed in the 1983 experiment, and I tried to duplicate my results, but without success. I was, however, able to get some useful information.

Attaching reflecting foil to the spindle of the 3-inch fiberglass disc top, I measured with a Heathkit reflecting tachometer a speed of just over 4000 rpm. Current output for my Wimshurst generator having 14-inch sectorless acrylic discs was 15 to 20 microamps with the meter leads connected across the two output terminals.

Spinning top construction

Because I don't have a lathe, I make do with the drill press in making the grinding wheel top shown in Figs. 11-10 and 11-11.

To put a hole in the shaft end of the top, I square the press table, having sandpaper on its surface and bring the end of the spindle down. See Fig. 11-12. This makes a square end in which to drill a hole. I mount a ⁷⁄₆₄-inch drill bit pointing up in the press vise, also shown in Fig. 11-12.

I use a high speed setting and *slowly* bring the spindle end down and drill a central hole about ¼ inch deep. I now flip the spindle ends, holding the round shank end, and screw a ¼–20 × ½-inch-long machine bolt tightly into the threaded hole in the shank shown in Fig. 11-13. I bring this down on sandpaper to establish the center and square the bolt head.

I mount a ¹⁄₁₆-inch drill in the vise as before, and drill a center hole about ¼ inch deep into the bolt head. Into this hole I press fit a compass point or ¹⁄₁₆-inch steel wire sharpened to a fine point; it projects about ¼ inch. With the point and spindle spinning true, I fix the point with a drop of super glue. I next install the grinding wheel (2 to 2½ inch diameter × ¼ inch thick), using a thin cardboard washer on each side of the wheel.

Finally, I true the grinding wheel edge with a wheel "dresser" so that it is concentric with the spindle. This completes the top's machine work. I cut a straight section of bamboo skewer 2 inches long. (Bags of skewers are sold in the kitchen/hardware sections of grocery stores.) This presses into the spindle hole, thus completing the top as shown in Fig. 11-11.

11-10
Two-inch-diameter aluminum oxide grinding wheel top. Point is positive, ball is negative.

Experiment procedure

In Fig. 11-14, the pointed electrode (¼ × 2 inch long aluminum rod) can be adjusted up or down on its acrylic base and likewise the ball electrode that is 2 to 3 inches in diameter. Insulated leads connect these electrodes to the Wimshurst generator. The top's support is set into a 2 × 12 × 18 inch slab of styrofoam.

I move the electrodes out to about ½ inch away from the rotor's edge. Then holding the end of the bamboo skewer, I give the top a quick twist with the fingers. It will precess on its rod support, and I now turn on the generator. The speed picks up and finally I move point and ball closer to the top to give a ¹⁄₁₆-inch air gap to the rotor. **Caution: Shield yourself, camera, and generator with acrylic plate around this setup to prevent any damage.** A soft base helps cushion tops in case they tumble off. I am not certain of the best polarity for the electrodes or whether Leyden jars are needed for optimal function. Recommended positions for the electrodes are shown in the upper portion of Fig. 11-14.

Larger Van de Graaff generators do give good results normally if their current output is above 10 microamps. The neon sign transformer (30 milliamps) might suffice as an option; however, it must be run through a voltage doubler or tripler to give a steady high-voltage dc output.

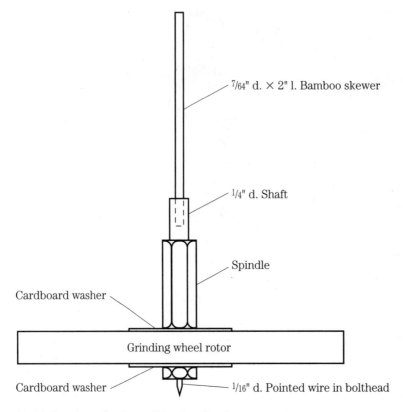

7/64" d. × 2" l. Bamboo skewer

1/4" d. Shaft

Spindle

Cardboard washer

Grinding wheel rotor

Cardboard washer

1/16" d. Pointed wire in bolthead

11-11 Grinding wheel top. Side view detail.

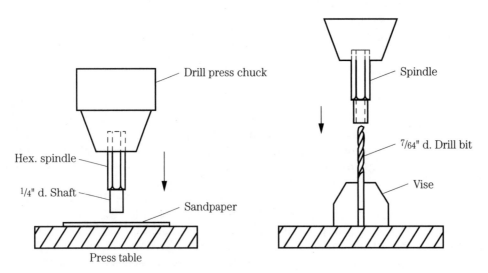

Drill press chuck

Hex. spindle

1/4" d. Shaft

Sandpaper

Press table

Spindle

7/64" d. Drill bit

Vise

11-12 Rotor shaft hole with bamboo skewer.

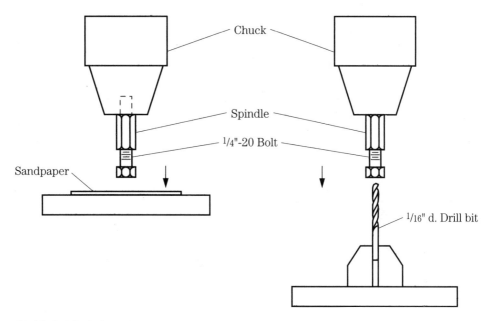

11-13 Bolt hole fit with needle point.

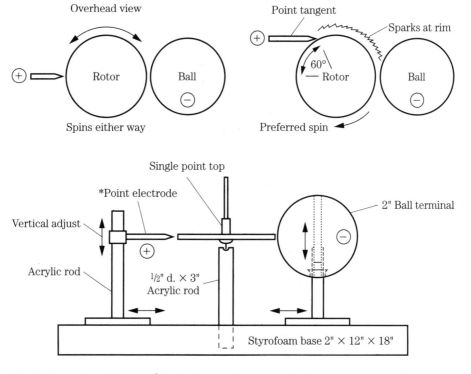

11-14 Experimental setup. *Protective acrylic cover not shown.

To test the conductive resistance of a rotor, I charge an electroscope leaf to the horizontal position and touch the rotor to its electrode. The leaf should drop no faster than 10 degrees in a 1-minute interval.

Rotor discs used to make tops require a high resistance. Too low a resistance for the rotor allows the charge to conduct and prevents any torque.

If you cannot locate small grinding wheels, then I recommend trying washers made from fiberglass, acrylic, fiber, phenolic, treated as described before.

I continue to experiment along these lines to reproduce this rare jumping top. In hindsight, it pays to keep good notes of experiments!

Supplies for experiments

G-C Red Glpt Insulating Varnish, catalog #10-9002, is sold through industrial electronic supply houses or may be ordered from Klaus Radio, Inc., 8400 N. Allen Road, Peoria, Illinois 61615.

Bulk aluminum oxide abrasives are sold through industrial supply houses for use with sandblasting equipment.

Why do aluminum oxide grinding wheels work so well?

I assume that the bulk resistance being so high, the relaxation time for charge distribution is long. This disequilibrium in charge may produce a torque by way of coulomb attraction and repulsion forces between the rotor and the ball terminal. If this is the correct explanation, then corona discharge streaming from the pointed electrode to the rotor, also called the "ion wind," plays a minor role in producing torque.

Experimenters should also try white aluminum oxide cup-shaped grinding wheels (used on tool and cutter grinders), as the white grade is the purist form and does have high resistance in bulk. Remember to shield the rotor for your own protection.

12
CHAPTER

Electrohorticulture

Electrohorticulture is the branch of high-voltage research that deals with the effects of electric fields on growth rate, quantity, and quality of vegetation. Experiments in this field have been traced to 1746 when Mr. Maimbray of Edinburgh, Scotland, studied the growth of myrtle trees. However, serious work on a large scale was forced to wait until the development of high-voltage influence machines, transformers, and better insulators.

In electrohorticulture, also called *electroculture*, electricity is applied to water sprayed on plants, to metal-coated seeds, or by way of air-to-earth electric fields using overhead wires. By far, most of the research concerns overhead wires charged with either high-voltage direct current, high-frequency alternating current (as with the Tesla coil transformer), and low-frequency periodic pulses of direct or alternating current at high potential.

From 1900 to 1915, considerable work was done with the influence machine. One such experimenter was Professor Selim Lemstrom of Helsingfors (at that time in Russia). In 1904, he published a book, *Electricity In Agriculture and Horticulture*, describing his theories, equipment, and results from many years of large-scale experiments. In 1903, he received U.S. patent number 720,711 describing his cylindrical influence machine (Fig. 12-1 and Table 12-1). This generator, driven by a ¹⁄₁₀ hp motor, employed a glass cylinder 30 cm in diameter by 40 cm in length. The generator was quite reliable over extended periods of operation.

Figure 12-2 shows Lemstrom's generator, located in a building for protection from the weather. Porcelain insulators on wood poles support 1½-mm galvanized wire running lengthwise and ½-mm wires laid crosswise 1 meter apart. The small wires are barbed with points. (See Table 12-1 for the results using this arrangement.) From 1910 to 1929, this research was reported in several scientific journals. The voltages were increased and wire grids were raised 8 to 12 feet above the ground for easier cultivation.

Some of the benefits claimed in these extended tests included yield increases from 21 to 65 percent, increased sugar content in fruits and vegetables, richer colors in flowers, and especially improved resistance to drought and diseases. Even though

Electricity in Agriculture and Horticulture, 1904

12-1 Lemstrom's patented generator.

the current-density requirement is tiny for setting up an effective electrical field (less than ½ milliamp per acre or 1.2×10^{-11} amps per square cm), this current is still 50,000 times the natural average air-earth current density!

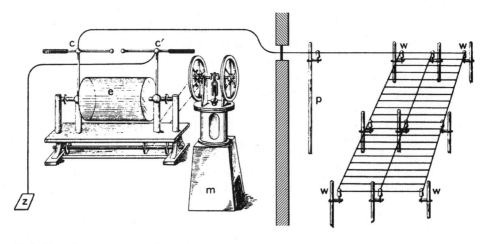

12-2 Charging setup for outdoor plots. *Electricity in Agriculture and Horticulture, 1904*

Table 12-1. Lemstrom's test results for several crops. *Electricity in Agriculture and Horticulture, 1904*

CHEMICAL ANALYSIS OF CROPS AT ATVIDABERG WITH AND WITHOUT ELECTRICAL TREATMENT.

[NOTE.—The increase per cent. of a substance is calculated after the formula $\frac{(a-b)100}{b}$, where a represents the quantity which was received under electric current, and b the quantity received outside the electrical current.]

SUBSTANCE.	RYE.			BARLEY.			OATS.			OATS separated fm. meslin.		
	Under electric current.	Outside electric current.	Incr'se per cent.	Under electric current.	Outside electric current.	Incr'se per cent.	Under electric current.	Outside electric current.	Incr'se per cent.	Under electric current.	Outside electric current.	Inc. or dec. percent
Mineral substance (ash)	2·128	2·440	—	2·979	2·950	—	3·712	3·739	—	3·899	4·107	—
Ether-extract (raw fat)	2·516	1·799	—	2·018	2·443	—	4·831	5·358	—	5·255	5·339	—
Cellulose	2·296	2·379	—	5·605	5·015	—	11·136	9·434	—	9·944	11·655	—
Nitrogen free extract	81·730	83·922	—	77·158	77·502	—	67·781	69·429	—	69·002	66·853	—
Raw proteid matter	11·270	9·460	19·13	12·240	12·090	12·41	12·540	12·040	4·15	11·900	12·040	-1·16
TOTAL	100	100	—	100	100	—	100	100	—	100	100	—
Per cent. of Nitrogen :												
Total nitrogen	1·803	1·513	19·17	1·958	1·935	1·19	2·007	1·927	4·15	1·904	1·927	-1·19
Amide nitrogen	0·218	0·110	—	0·094	0·061	—	0·093	0·112	—	0·086	0·131	—
Albuminoid nitrogen	*1·427*	*1·249*	14·25	*1·567*	*1·599*	2·00	*1·795*	*1·660*	8·13	*1·724*	*1·661*	3·8
Total digestible nitrogen	1·645	1·359	21·05	1·661	1·660	0·06	1·888	1·772	6·55	1·792	1·792	0·0
Non-digestible nuclein	0·148	0·154	—	0·297	0·285	—	0·119	0·155	—	0·112	0·135	—
Digestibility of nitrogen	91·8	89·8	—	84·8	85·8	—	94·0	91·1	—	94·1	93·0	—
In 100 parts of Nitrogenous Matter:												
Amides	12·7	7·2	—	4·8	3·2	—	4·6	5·8	—	3·6	6·8	—
Digestible protein	79·1	82·6	—	80·0	82·6	—	89·4	86·1	—	90·5	86·2	—
Non-digestible protein	8·2	10·2	—	15·2	14·2	—	6·0	8·1	—	5·9	7·0	—

From this analysis it follows that the electrical air current has produced an increase of the proteid matter in the rye of 19· per cent., in the total quantity of nitrogen 19·2 per cent., in the albumen of 14·3 per cent., and in the digestible nitrogen of 21·1 per cent. In the barley the electric air current only produced in the raw proteid matter an increase of 12·4 per cent., in the total quantity of nitrogen 1·2 per cent. In the oats, on the contrary, it has produced an increase of 4·2 per cent. of proteid matter, of 8·1 per cent. in the albumen, and of 6·6 per cent. in the digestible nitrogen ; whereas in the oats separated from the meslin an increase of albumen of only 3·8 per cent. was produced. All crops under the electrical treatment had been improved in quality to a considerable extent. For rye this improvement can be taken at 20 per cent., for barley at 12 per cent., and for oats at 10 to 12 per cent.

The above analysis was made in the Agriculture Economic Laboratory at Helsingfors (under the direction of Prof. Arthur Rindell) by candidate of philosophy Mrs. Lilly Wendt.

After 1929, research rapidly declined in the United States and Britain. It is not clear whether the reduced research resulted from the economic depression or because of fears about overproduction. A study of the history of this unusual aspect of agriculture would itself be most interesting.

If you are interested in this area of electrification, study the patents issued—for example, U.S. patent number 1,268,949 of June 1918 to Reginald Fessenden and U.S. patent number 1,952,588 of March 1934 to Kenneth Golden.

Our existing "fence-charger" technology, so popular on farms today, might be modified for some electroculture experiments. However, be careful to avoid causing interference to the radio bands—this is an FCC violation that carries stiff fines.

13
CHAPTER

Electrotherapy

The branch of medical treatment that uses electric fields for diagnosis and cure of disease is called *electrotherapeutics*. Electrotherapeutics involves the study of electrobiology and especially electrophysiology. Just as plants respond to electric fields, so do animals and people.

The first recorded mention of electrical treatment appears during the first century A.D. when the electric torpedo fish produced numbness to the legs and feet, relieving pain from gout. During the 1700s, electrical experimenters carefully studied the torpedo fish, trying to find the source of its electricity. As electrical science developed during the eighteenth century, more experiments were attempted to treat diseases in human subjects—first with static electricity and later with high-frequency currents. In the 1800s, both galvanic batteries and Faraday's transformers were used for specific types of electrotherapy.

The explosive growth of electrical science from the 1890s to the 1920s provided many more instruments and techniques for the electrotherapeutic physician. Both in Europe and North America, science journals dealing with electrotherapy and radiology appeared, and professional associations were formed. Most physicians who used electrotherapeutics were from the regular allopathic school. When drugs failed to help a patient, then electrical methods were used.

Shown in Fig. 13-1 is a popular electrostatic generator most often used by American physicians. Note the tray containing electrodes and Leyden jars.

Figure 13-2 shows one of the many setups for using the Holtz or Wimshurst machines. Figure 13-3 shows one of the generator advertisements from the turn of the twentieth century.

During the 1920s, an increasing interest in alternating current as a new technology, especially at radio frequencies, resulted in electrotherapeutic transformers using Tesla and Oudin coils. At this time, several researchers were beginning to study the electrical aspects of physiology.

One result from these experiments was the introduction in 1929 of the electroencephalograph (EEG), which measures the electrical activity of the brain. However, in spite of considerable interest in the electrical aspects of biology, the use of

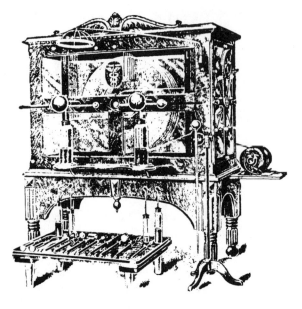

13-1
Electrotherapy generator. *Essentials of Modern Electro-Therapeutics*, Strong, 1908

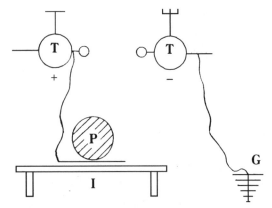

13-2
Illustration of usage. *Essentials of Modern Electro-Therapeutics*, Strong, 1908

T T, Terminals of Static Machine.
I, Insulating Platform.
P, Patient.
G, Ground Connection.

electrotherapy as a recognized branch of medicine declined, especially in America. After some time, even discussing electrobiology with physicians was considered blasphemy. In spite of the well-developed methods, equipment, and efforts expended, the term *electrotherapy* is rarely listed in modern English encyclopedias.

Explaining how electrotherapy disappeared requires a careful study of the powerful political and economic forces that developed during the early 1900s. The pivotal year was 1910, during which the Carnegie Foundation issued a scathing exposé on the deplorable conditions of American medical schools; this was the Flexner Report, named after its chief investigator, Abraham Flexner. The result was that about half

13-3 A typical electrotherapy advertisement. *Journal of the Rontgen Society,* 1905

of these schools closed within the year. While the report served as a catalyst for improving the standards in medical education and rehabilitation in medical practices, the resulting homogenization and sterilization process led to a great narrowing of options. Thus a monopoly developed favoring allopathy as opposed to homeopathy. For most of the twentieth century, "drugs as medicine" has been the favored paradigm in American medical practice. What was virtually outlawed included homeopaths, chiropractors, osteopaths, electrotherapy, mental healers, and spiritual healers. Fortunately, as a result of public demand within the last decade, alternative therapy has blossomed once again.

In spite of the general loss of this innovation in medical practice, several vestiges still remain in Western culture. In addition to the development of the EEG is electric stimulation for muscle toning, the pacemaker for heart patients, diathermy and microwave heating, nuclear magnetic resonance, and last, the most uncivilized of electrotherapeutics, electroshock "therapy."

In other parts of the world, especially in eastern Europe and Asia, there has been a greater toleration for alternatives to the drugs-as-medicine monopoly that rules in the West. As a result, many of the developments in electrotherapy and diagnosis are fortunately still being used overseas.

It is interesting to note how the Western medical establishment has responded to the Far Eastern concept of health. Western physicians once denounced the ancient practice of acupuncture as merely having a "placebo effect." But its use for pain relief during advanced cases of surgery dispelled this superficial explanation.

The issue behind both acupuncture and electrotherapy is the quantification versus the qualification of health. Orthodox Western medical practice treats the body as a mechanism but neglects studies of energy flow in the body. In recent years, there has been an amalgamation of disciplines that treat energy flow and health. Electroacupuncture, for example, stimulates specific points on the body with electricity to induce changes in bioenergy flow or to remove a blockage to this flow.

Ancient Eastern concepts of health are beginning to meet with Western methods of practice, so we can expect to see a more holistic approach to bodily health, with greater attention paid to the feeling and quality of life, reflected through the use of preventative treatment.

The present crisis in orthodox Western medical practice, involving both chronic and contagious diseases, creates a public demand for access to alternatives. Employing medically unconventional methods should also give rise to a tolerant review of the concepts in electrotherapy, in light of the recent advances in biophysics.

14
CHAPTER

High-voltage humans

To counter the authoritative claims denying the important electrical functions in the human body, I provide two cases of anomalous electrophysiology. Through the centuries there have been several eyewitness accounts of this phenomenon. One account was mentioned in the *Annals of Electricity, Magnetism and Chemistry* (vol. II, 1838).

Figure 14-1 describes the case of an "electrified woman." The article finished by saying that the lady was about 30 years old, of a delicate constitution and nervous temperament, had sedentary habits, and seldom had been confined to her bed by sickness.

My second and more recent example (Fig. 14-2) was found in *Science and Invention* (1920).

The Clinton convicts were also mentioned in the *New York Times*, Monday, April 5, 1920, page 1.

Very recently, there have been reports from the New China News Agency of high-voltage humans. Dr. Chang Gee Sun of Beijing has been actively studying this area of electrophysiology. Although there might be living examples in North America, the science journals as a rule ignore such difficult-to-explain anomalies.

If you are interested in the recent innovations in electrobiology, read *The Body Electric* by Robert Becker and Gary Selden (1985). If you wish to research the interesting history of electrotherapy, visit the Bakken Library and Museum in Minneapolis, Minnesota. The electrical exhibits are primarily from the years 1700 to 1900, and include many of the electrostatic generators mentioned in this book.

A lady of great respectability, during the evening of the 25th of January, 1837, the time when the aurora occurred, became suddenly and unconsciously charged with electricity, and she gave the first exhibition of this power in passing her hand over the face of her brother, when, to the astonishment of both, vivid electrical sparks passed to it from the end of each finger.

The fact was immediately mentioned, but the company were so sceptical that each in succession required for conviction, both to see and feel the spark. On entering the room soon afterward, the combined testimony of the company was insufficient to convince me of the fact until a spark, three fourths of an inch long, passed from the lady's knuckle to my nose causing an involuntary recoil. This power continued with augmented force from the 25th of January to the last of February, when it began to decline, and became extinct by the middle of May.

The quantity of electricity manifested during some days was much more than on others, and different hours were often marked by a like variableness; but it is believed, that under favorable circumstances, from the 25th of January to the 1st of the following April, there was no time when the lady was incapable of yielding electrical sparks.

The most prominent circumstances which appeared to add to her electrical power, were an atmosphere of about 80° Fah., moderate exercise, tranquillity of mind, and social enjoyment; these, severally or combined, added to her productive power, while the reverse diminished it precisely in the same ratio. Of these, a high temperature evidently had the greatest effect, while the excitement diminished as the mercury sunk, and disappeared before it reached zero. The lady thinks fear alone would produce the same effect by its check on the vital action.

We had no evidence that the barometrical condition of the atmosphere exerted any influence, and the result was precisely the same whether it were humid or arid.

It is not strange that the lady suffered a severe mental perturbation from the visitation of a power so unexpected and undesired, in addition to the vexation arising from her involuntarily giving sparks to every conducting body, that came within the sphere of her electrical influence; for whatever of the iron stove or its appurtenances, or the metallic utensils of her work box, such as needles, scissors, knife, pencil, &c. &c., she had occasion to lay her hands upon, first received a spark, producing a consequent twinge at the point of contact.

The imperfection of her insulator is to be regretted, as it was only the common Turkey carpet of her parlor, and it could sustain an electrical intensity only equal to giving sparks one and a half inch long ; these were, however, amply sufficient to satisfy the the most sceptical observer, of the existence in or about her system, of an active power that furnished an

14-1 A high-voltage human. *Annals of Electricity, vol. 2, 1838*

uninterrupted flow of the electrical fluid, of the amount of which, perhaps the reader may obtain a very definite idea by reflecting upon the following experiments. When her finger was brought within one sixteenth of an inch of a metallic body, a spark that was heard, seen, and felt, passed every second. When she was seated with her feet on the stove-hearth (of iron) engaged with her books, with no motion but that of breathing and the turning of leaves, then three or more sparks per minute would pass to the stove, notwithstanding the insulation of her shoes and silk hosiery. Indeed, her easy chair was no protection from these inconveniences, for this subtle agent would often find its way through the stuffing and covering of its arms to its steel frame work. In a few moments she could charge other persons insulated like herself, thus enabling the first individual to pass it on to a second, and the second to a third.

When most favorably circumstanced, four sparks per minute, of one inch and a half, would pass from the end of her finger to a brass ball on the stove; these were quite brilliant, distinctly seen and heard in any part of a large room, and sharply felt when they passed to another person. In order further to test the strength of this measure, it was passed to the balls by four persons forming a line; this, however, evidently diminished its intensity, yet the spark was bright.*

The foregoing experiments, and others of a similar kind, were indefinitely repeated, we safely say hundreds of times, and to those who witnessed the exhibitions they were perfectly satisfactory, as much so as if they had been produced by an electrical machine and the electricity accumulated in a battery.

The lady had no internal evidence of this faculty, a faculty sui generis; it was manifest to her only in the phenomena of its leaving her by sparks, and its dissipation was imperceptible, while walking in her room or seated in a common chair, even after the intensity had previously arrived at the point, of affording one and a half inch sparks.

Neither the lady's hair or silk, so far as was noticed, was ever in a state of divergence; but without doubt this was owing to her dress being thick and heavy, and to her hair having been laid smooth at her toilet and firmly fixed before she appeared upon her insulator.

As this case advanced, and supposing the electricity to have resulted from the friction of her silk, I directed (after a few days) an entire change of my patient's apparel, believing that the substitution of one of cotton, flannel, &c., would relieve her from her electrical inconveniences,* and at the same time a sister, then staying with her, by my request, assumed her dress or a precisely similar one; but in both instances the experiment was an entire failure, for it neither abated the

* It is greatly to be regretted that the spark had not been received into a Leyden bottle until it would accumulate no longer, and then transferred to a line of persons to receive the shock.—ED.

14-1 Continued.

intensity of the electrical excitement in the former instance, or produced it in the latter.

My next conjecture was, that the electricity resulted from the friction of her flannels on the surface, but this suggestion was soon destroyed when at my next visit I found my patient, although in a free perspiration, still highly charged with the electrical excitement. And now if it is difficult to believe that this is a product of the animal system, it is hoped that the sceptics will tell us from whence it came.

14-1 Continued.

Poisoned Convicts Become "Electrified"!

One of the First of the Phenomena Noted In the Case of "Botulinus Poisoning," Caused By Eating Decayed Canned Salmon, Was That the Body of the Patient Had Become Highly Electrified. He Was Unable, for Example, to Throw a Piece of Paper In the Waste Basket, the High Electric Charge In His Body Attracting the Paper to His Hand.

AS per schedule, the case of the thirty-four convicts at Clinton Prison, Dannemora, N. Y., who became poisoned by eating canned salmon, and thereafter develop remarkable electrified propensities, was fed to us for several days by the ever-busy newspapers, under the captions of "human magnets" and what not. The facts in the glaring case are here presented for the first time.

The following details relative to *botulinus poisoning*, which took place at this institution, February 20th, 1920, are cited in a letter which we have received from the chief physician at Clinton Prison, Dr. Julius B. Ransom. Dr. Ransom says:

Among Other Things the Electrified Patient Was Able to Move a Suspended Steel Tape Measure and Also to Attract the Filament Of An Incandescent Lamp Towards the Side Of the Globe.

"Dr. Rosneau, of Harvard University, did not make any investigations of the *electric phenomena* and only came into the case with reference to the *botulinus poison*, as it was a rather large group of cases and opportunities for study were unusually good. Of course the newspaper reports were garbled and exaggerated as they usually are when they attempt to report scientific matters. The newspaper accounts were taken from a report made by myself to the Superintendent of Prisons, setting forth the history and development of 34 cases of *botulinus poisoning, due to the eating of canned salmon.*

"During the course of these cases it was discovered by accident that peculiar static electric power had developed in the patients. It was discovered in this manner. One of the patients who was convalescing crumpled up a piece of paper, I imagine in both hands, and attempted to throw it in a waste basket; it absolutely refused to leave his hand. From this time on experiments were made, and the matter was reported to me, and I found that *every* case of *botulinus poisoning* developed this strange power, and that neither the attendants nor nurses associated with them had any such power. All sorts of experiments have been tried and it was found to be a constant condition; that is, that this peculiar power of creating a magnetized (electrified) field by rubbing the hands together, which puts them in circuit, will electrify different objects, so that they will retain that electrification for many hours. For instance forms of paper, such as newspapers, and ordinary correspondence paper when electrified by these patients and thrown against the wall adhered and clung to any object for many hours. By again rubbing the hands together and rubbing the electric light bulb the filament will begin to vibrate very rapidly and follow the motions of the hands, and attach themselves to the side of the bulb with a good deal of sparking at the base of the filament. The compass needle of a surveyor's instrument can be rotated with any piece of paper electrified by these patients. A steel tape suspended, will feel the magnetic field in a remarkable manner and sway from side to side.

"What relation there can be between the botulinus toxin and this phenomena of course is difficult to identify; it has been suggested that it is the *dryness of the skin* which prevents the ordinary passing out or dissipation of the electric currents from the body; *but the patient submerged in bath tub performs the same phenomena as when clothed!* The ability to electrify is propor-

Another Phase of the Electrified Paper Phenomenon, Due to the Patient's High Potential Electric Charge Occasioned by Botulinus Poisoning. A Sheet of Paper Electrified by the Patient Would Remain Against the Wall for Hours. He Was Also Able to Move the Compass Needle of a Surveyor's Instrument.

tioned to the severity of the disease; as the patient convalesces he gradually loses this power and when quite well loses it altogether.

"I might mention further that all these cases were ataxic and developed peculiar reflexes. Many of them were almost entirely blind and had paralysis of the upper lid "Ptosis." Of course, in *botulinus poisoning* the nervous system is about the first to suffer; one thing is quite clear, therefore, static manifestation is closely linked with the disturbance of the central nervous system and represents, no doubt, simply a much higher degree of static storage in the body than is usual."

Electricians Argued That If the Patient Was Placed In a Tub Full of Water, That the Charge Would Disappear, But Strangest of All It Did No Such Thing—and the Patient Was Still Able to Attract a Steel Tape Measure or Other Object By Electro-static Attraction.

15
CHAPTER

Cold light

By *cold light*, I mean the production of light without necessarily the evolution of heat. Work in this field has been inspired by the many examples provided in nature including fireflies, glowworms, auroras, St. Elmo's fire, and earthquake lights.

Since the incandescent light bulb has an energy efficiency of about 8 percent and the fluorescent bulb an efficiency of 15 to 20 percent, any improvements would, of course, greatly conserve our energy resources (and reduce air conditioning loads). Even though cold light has been closely studied for over 200 years, the subject is one of the least-understood physical phenomena.

The auroral light

The largest and most spectacular display of nature's cold lights are the *aurora borealis* and *aurora australis*. The luminous flickering red, yellow, blue, and green lights are seen near the northern and southern polar regions. The relationships between electrical activity and the auroras were most carefully studied near the turn of the twentieth century by Professor Selim Lemstrom of Finland, who in 1883 reported his experimental results (Fig. 15-1).

Lemstrom succeeded in artificially producing the rare low-level aurora (where streamers reach to the ground) by arranging a ground-insulated flat spiral of wire that covered 900 square meters on top of a hill near Kultala, Finland. A single beam of cold light formed over this installation and extended to a height of 400 feet; the spectroscope showed a greenish-yellow spectrum, with a 5569-angstrom wavelength of varying intensity. This is the only known experiment that successfully reproduced the properties of the aurora on a large scale.

In an age of exotic, expensive "high" technology, we should marvel at how much Professor Lemstrom accomplished by the simplest of methods, intuitively directed. His experiments show a marvelous, elegant frugality.

On a smaller scale, the Norwegian scientist Kristian Birkeland was able to simulate auroral light. He used a positively electrified copper sphere in a vacuum chamber,

A VERTICAL SHEAF OF LIGHT OBSERVED, DURING A DISPLAY OF THE NORTHERN LIGHTS, ABOVE A SYSTEM OF WIRES ON THE TOP OF PIETARINTUNTURI, NEAR KULTALA, FINNISH LAPLAND.   (Reproduced from *La Nature*.)

15-1 A vertical sheaf of light observed over Lemstrom's apparatus. *Science*, n.s., vol. 4, 1884

with a negatively charged disc at the side of the box emitting charged particles. The sphere represented the earth in space and the disc functioned as the sun.

His main work, *On the Cause of Magnetic Storms* (1908), features photos of his experiments (Figs. 15-2 through 15-4). In Birkeland's experiments, power was supplied at 15,000 Vdc at 500 milliamps. The sphere diameter varied from 8 to 40 cm, and the chamber was evacuated down to a few hundredths of a millimeter.

Earthquake lights

Earthquake lights, which can appear before, during, and after earthquakes, have been recorded since ancient times. The lights take the form of fireballs, streamers, rays of light, flames, or glowing mists of various colors. Several good photographs of this phenomenon have been taken.

The Japanese and Chinese have been careful to preserve good records because the lights can help give short-term predictions for quakes. In one such occasion, on November 2, 1931, at South Hyuga, Japan, the lights appeared on the horizon as divergent rays of blue light. Such a phenomenon can last from several seconds to over a minute.

15-2
Birkeland's laboratory auroras.
On the Cause of Magnetic Storms, Kristian Birkeland, 1908

15-3 Birkeland's aurora.

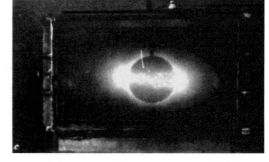

15-4 Birkeland's aurora continued.

Because of the multiple manifestations of earthquake lights, they cannot all be explained away as power outages or ruptured gas lines. An excellent discussion of these lights is included in *Lightning, Auroras, Nocturnal Lights,and Related Luminous Phenomena* by William R. Corliss (1982).

Possibly related to earthquake lights, but on a smaller scale, is the phenomenon called *triboluminescence*, that is, light produced by impact, friction, or rubbing. In addition to smoky, amethyst, or rose quartz, and flint, sugar crystals show brief flashes of light when they are struck or fractured.

A simple, but "illuminating," experiment follows: Allow your eyes to become adjusted to a completely dark room for about 20 minutes. Now chew Wint-O-Green or Pep-O-Mint Lifesavers mints and have a friend observe (or with a mirror observe) the bluish flashes of light in your mouth as you break up the candy. Loaf sugar and rock candy also produce this effect.

Professor Linda Sweeting, from the chemistry department at Towson State University in Towson, Maryland, has been studying the spectrum of "mouth lightning." Sweeting relates the color to the nitrogen in the air and the separation of electric charges by fracturing.

By extrapolation, you can visualize how, on a large scale, triboluminescent rocks such as quartz look when they are fractured or rubbed together during a quake or in a rock avalanche. The tons of material and surface area involved should produce quite a light display. Released flammable gases, such as methane, would produce jets and sheets of flame, quite distinct from cold light.

This branch of cold light, if seriously studied, could help with short-term earthquake prediction and also with locating minerals in the earth's crust by spectral analysis. If a large number of people in earthquake-prone areas were educated on this subject, then it might be possible to classify these light forms.

Invisible phosphorescence

Invisible phosphorescence is the least-studied form of cold light. This cold light was discovered in 1899 by the incomparable Dr. Gustave Le Bon and described in his work, *The Evolution of Forces* (1908). When phosphorescent compounds such as calcium sulphide are exposed once to light then kept in the dark, the materials continue to radiate, sometimes for months. Although these radiations are not visible to the eye, they can be recorded on film. A related form of light is shown in Fig. 15-5.

Figure 15-5 is a slight variation of mine in which the 14-inch Wimshurst machine has both a large 2-inch-diameter wood ball terminal, which is painted with luminous paint, and an uncoated ¾-inch steel ball terminal.

When you take a photo, you can only see several sparks over the 6-second interval. Imagine my surprise after developing the film to note the continuous bright glow. Evidently streams of paint particles are dislodged by the discharge, but they radiate light, mostly at a wavelength only captured on film (Fig. 15-6). Le Bon was even able to take photographs through opaque bodies using these invisible radiations!

It would be most profitable to continue the research work of a number of pioneers. Those who were heavily involved in the study of phosphorescence and fluorescence include Edmond Becquerel (late 1850s), William Crookes (1880s), Gustave Le Bon (1890s), Johann Puluj (1890s), and Hermann Ebert (1890s).

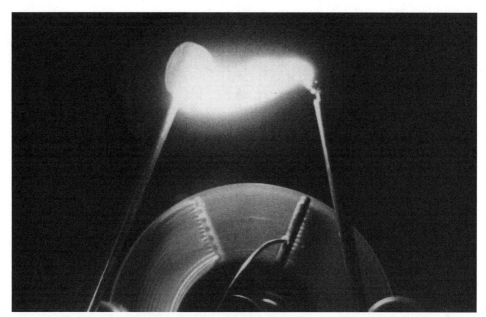

15-5 Invisible electro-phosphorescence. Film: Black & white ASA 400, F/1.7, 6 seconds.

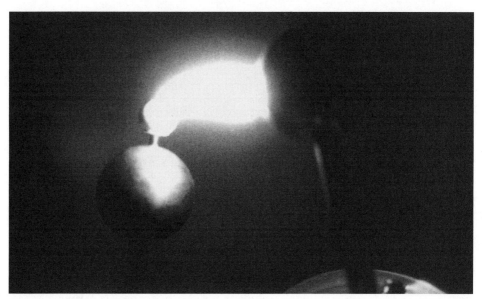

15-6 Invisible electro-fluorescence from a sulfur-coated metal sphere. Film: Black & white ASA 400, f/1.7, 6 seconds.

The "electric egg"

Experimentally, the nature of cold light in controlled atmospheres, such as rarefied gases, might be studied by constructing the "electric egg" (Fig. 15-7).

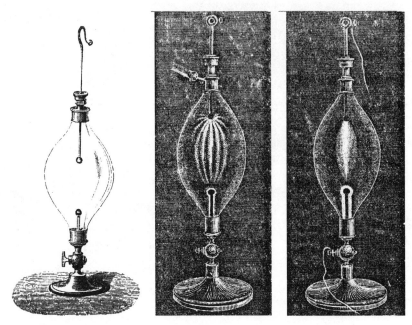

15-7 The "electric egg." *Electricity and Magnetism*, A. Guillemin, 1891

The egg-shaped glass chamber has two metal rods. Terminate each rod with a ball, which can be brought close or separated as desired. Remove the egg from its stand and join it to a vacuum pump. Control the pump's flow with the stopcock at the base of the chamber.

At a vacuum pressure of 60 mm, electric discharges appear, as shown in Fig. 15-7. At approximately 3 mm of mercury pressure, the discharge appears as a red-tinted luminous sheaf issuing from the positively charged ball. The negatively charged ball and rod are enveloped by a layer of bluish-purple light. A water aspirator or a discarded refrigerator compressor can provide an adequate vacuum pressure for many cold light experiments.

Phosphorescent lamps

During 1896 and 1897, several articles appeared in science journals on the phosphorescent lamp developed by Austrian researcher Johann Puluj. His lamp (Fig. 15-8) was shaped like an Edison incandescent lamp, but the wires extended through the bulb rather than through the socket. Both wires were made from aluminum. The negative pole (cathode) ended in an aluminum reflector-shaped disc. Hanging from the apex of the globe was a small square sheet of mica. The mica surface facing the reflector was painted with calcium sulphide (key ingredient in luminous paint). Radiant electricity converged from the disc onto the painted mica anode. The anode glowed with a brilliant phosphorescent (cold) light.

Caution: When working with vacuum bulbs do not use a very high voltage or a very low vacuum pressure. If you use these, X-rays, which result from the sudden stoppage

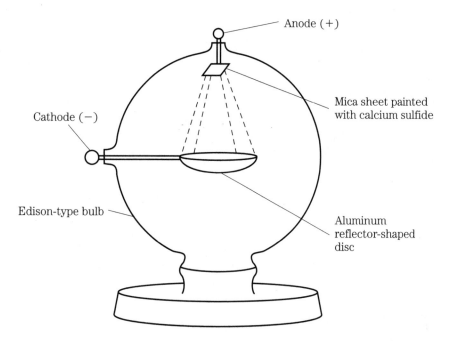

Anode (+)

Mica sheet painted
with calcium sulfide

Cathode (−)

Edison-type bulb

Aluminum
reflector-shaped
disc

15-8 Puluj's electrophosphorescent lamp.

of high-speed electrons, will be produced. Moderate pressures and voltages will keep the lamps cool and reduce deterioration of the phosphorescent compounds.

Researchers wanting to learn vacuum techniques should consult *An Experimenter's Introduction to Vacuum Technology* by Steve Hansen. This up-to-date booklet provides the details on practical hands-on methods and projects related to cold light studies.

A number of vacuum discharge tubes are available and are manufactured by Electro-Technic Products, Inc. of Chicago, Illinois. Figure 15-9 shows one of their glass tubes, which is 40 cm long and has a concave cathode.

One problem encountered when building homemade vacuum tubes and chambers is how to design a vacuum seal that allows electrodes to be slid in or out as well as rotated without having to dismantle the tube. The following seal is adapted from a sketch by N. Fuschillo, which was published in the *American Journal of Science*, vol. 26, 1958. Using cast acrylic tubing or a polycarbonate bell jar, this seal would be good down to one millitorr (one micron of mercury) vacuum. It is constructed of plumbing parts commonly available at hardware stores.

Figure 15-10 shows the end fitting for the 1½-inch acrylic tube; it is a 1¼-inch PVC pipe plug, which is turned on the lathe to remove threads and two grooves are cut to receive O-rings. The compression union seals the PVC tube using one brass ring, but the other ring is replaced with a #1 holed soft rubber stopper, which compresses against the sliding ⁵⁄₃₂-inch brass rod with terminal installed.

The union cap on the right side controls the seal pressure on the brass rod. A small amount of vacuum grease on the seal parts helps prevent any leaks.

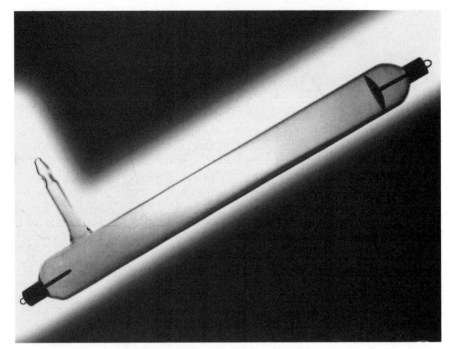

15-9 Vacuum discharge tube #CD 400. Electro-Technic Products, Inc. Chicago, Ill.

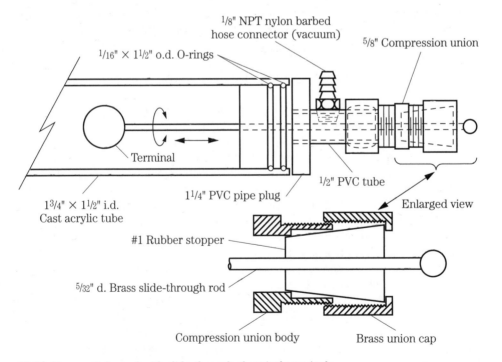

15-10 Vacuum tube seal with slide-through electrical terminals.

16
CHAPTER

Miscellaneous experiments

A hair-raising experiment

In Fig. 16-1, our radiant beauty, Diana, is charged to about (+)300,000 volts. She holds a ¼-inch-diameter aluminum rod rounded at both ends and brings one end near the positive terminal, while the negative terminal is grounded with a jumper wire. Your subject's hair must be clean and dry. No hair sprays should be used; they can be flammable.

Warning! Leyden jars must be removed for safety. The platform on which our model stands is a thick wood slab with legs made of PVC plastic pipe 2½ inches diameter × 4 inches long. This insulated platform is placed on a large rubber mat to prevent sparkovers. With this arrangement, the insulated person will reach the potential of the terminal.

At this energy level, all exposed human hair stands at attention, and the sensation feels "prickly." A higher potential, 500,000 volts, for example, at the same current, would cause an uncomfortable stinging sensation.

Extrapolating from this experiment, consider the experiences of a bird flying from earth to a high-voltage power line. Its body potential will start at "neutral ground," but as it nears the wire, its electrical potential rises. When it touches the wire, it is at the wire's potential, normally 3000 to 7000 volts (alternating or direct current doesn't affect the result). The bird feels nothing at contact. The electrified wire modifies the properties of the surrounding space.

Besides this hair-raising experiment, you can map out electric fields around the generator by blowing bubbles from the insulated stand. Bubbles usually fly to the walls or ceiling along peculiar paths. Figure 16-2 illustrates this experiment. Kristine holds a battery-powered bubble blower with a cord magnetically joined with the Van de Graaff's terminal. The other end of the cord clips to a 4-inch square of aluminum foil in good contact with her skin. Generator speed must be very slow and the stand well-insulated.

16-1 Author's version of a room-temperature SUPER conductor.

The levitating rocket

The next novel experiment involves levitating lightweight objects in space with electric forces. A number of patents have been granted for scientific toys that use this principle. A charging wand, operated by friction, normally provides the electric

16-2 Mapping electric fields with charged bubbles.

charge. If you like to make toys that illustrate scientific principles, study U.S. patents 2,018,585 October 1935; 3,497,994 March 1970; and 4,109,413 August 1978. A more "advanced level of play" makes use of the influence machine. The writings of Dr. Gustave Le Bon feature his description of the electrified "rocket."

Figure 16-3 shows the rocket levitating in space between a positive-charged metal sphere and a grounded, pointed metal probe. A painless way to launch the rocket is to charge the end of a ½-inch-diameter PVC pipe by touching it against the charged sphere (Leyden jars removed). Now place Le Bon's rocket, cut from very thin aluminum foil, crosswise near the end of the charged pipe. Place the rocket at the midpoint between the sphere and the probe and slowly roll the pipe. This action dislodges the rocket, and it will normally remain suspended in space as if it were held by invisible springs.

16-3
Spinning "rocket" below
3-inch sphere.

Various aluminum foil satellites

Figure 16-4 shows several shapes cut from very thin aluminum foil for use with the Wimshurst or Van de Graaff. Other materials including gold leaf, foam plastic, feathers, cork, and pith should be tried. Two small satellites may be supported in place of a larger single satellite!

Figure 16-5 shows a narrow wedge-shaped aluminum foil satellite with quarter twist that is spinning and hovering below a large 3-inch negative terminal with a metal plate below. Slanting the terminal handle downward slightly keeps the satellite from wandering away from the ball toward the aluminum tube. It helps to ground the opposite terminal for these experiments. **Warning:** Always remove Leyden jars for these experiments to avoid hazardous shocks! Remove insulated shoes, and let your feet be grounded so that you do not build up charges standing close to the terminal. Increasing generator speed moves the satellite farther away from the terminal, whereas lowering the speed brings the satellite nearer the terminal.

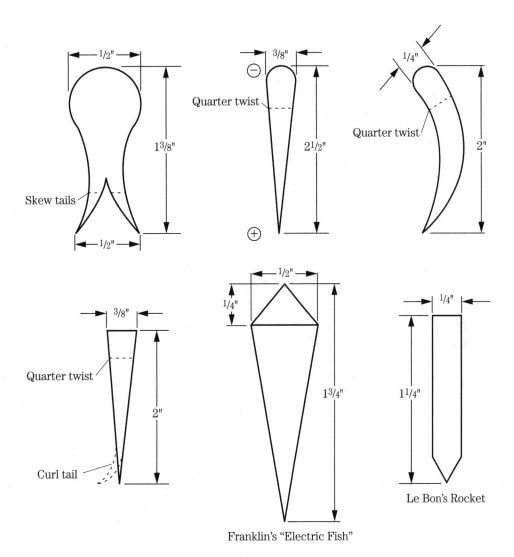

16-4 Various shapes for aluminum foil satellites.

As a rule, the narrow pointed end of the foil should face a positive terminal and the more rounded (obtuse) end should face a negative terminal. Otherwise, the satellite will be thrown out of the field. The space below the terminal may be open (without chairs or tables which can distort the electric field). A bare floor without rugs or a grounded metal plate below enhances the effect. Curling or skewing the tails, like a propeller, causes rapid spinning, and when properly shaped, produces a flight similar to the hummingbird.

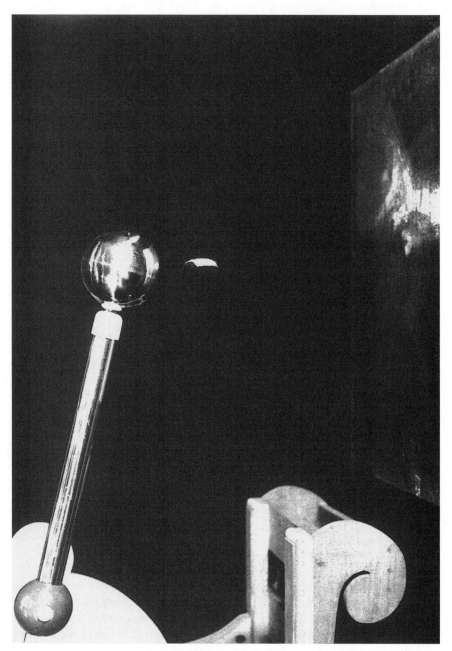

16-5 A spinning and levitating aluminum foil satellite.

17
CHAPTER

Electroaerodynamics

In *electroaerodynamics*, electric charges are applied to high-speed vehicles, such as supersonic aircraft, for the purpose of reducing air drag or eliminating sonic booms. In one method, high-speed ions are projected forward from the leading edges of the craft. This corona discharge propagates "upstream," and repels motionless air molecules away from the aircraft's oncoming surfaces.

In effect, this system should keep the relative motion between the vehicle and the adjacent air molecules so low that a shock wave cannot be mechanically produced. Of course, electroaerodynamics, by reducing frictional drag, could greatly conserve fuel resources and reduce air pollution. It has been estimated that a cut in drag of 1 percent can increase an aircraft's payload by 10 percent.

In the early 1960s, Horace C. Dudley received U.S. patent number 3,095,167 on his means for increasing the altitude of his electrified model rockets. However, his results need to be conclusively verified by other model-rocket enthusiasts.

In March 1968, an article was written on the experiments of Dr. Gustav Andrew and Maurice Cahn, who worked at Northrop's Norair Division in Hawthorne, California. In a 5-inch supersonic wind tunnel, they were able to propagate a corona glow 6 to 8 inches upstream from a ½-inch-diameter charged probe. This experiment proved that a flow of ions could charge the air molecules that precede a supersonic aircraft. If you are interested in attempting this experiment, you should insulate the throat of the wind tunnel well so that the charge is not lost to the walls. In addition to the high voltage, a substantial current will be needed to affect a large volume of air molecules. The design of the electrodes and of the leading edges is a crucial aspect because it facilitates uniform spraying of ions in a forward direction. An efficient influence generator would be quite adequate for use in preliminary small-scale wind-tunnel tests.

In the late 1970s, Anatoly Klimov at Moscow Radio-Technological Institute and colleagues at the Ioffe Institute in St. Petersburg were studying how shock waves behave in ionized gases. In one experiment, a walnut-sized steel sphere was shot at 1 kilometer per second through a tube filled with argon gas at low pressure. One section of the tube was ionized. What was captured on film showed the shock wave standing twice as far from the sphere as it would in a normal, un-ionized gas. Calculations showed a reduction in drag by 30%! Work has continued up to the present time, especially in Russia, Great Britain, and the United States, on understanding the very rich and complex physics of plasma dynamics in relation to aerodynamic surfaces for subsonic, supersonic, and hypersonic speeds. Recent experiments have shown that these "nonequilibrium" plasmas involve not just one dimension but two- and three-dimensional flow because of how the ions and electrons interact with each other in relation to leading edges. The conclusion must be that more than a simple heating effect is involved; therefore, far less electric power is needed to influence airflow. From an electrostatics viewpoint, very high speed microvortices could form, with ions and electrons forming "rotors" driven by the electric field, just as in an electrostatic motor. These movements would strongly influence air pressure ahead of the vehicle. In America, new experiments are being carried on at Arnold Engineering Development Center at Arnold Air Force Base in Tennessee. (See related article "Plasma Magic," by Justin Mullins, in *New Scientist,* vol. 28, October 2000, pp. 26–9.)

Above all, do not be dissuaded from experimenting. A mathematical disproof that indicates the power requirements are excessive for practical purposes cannot cover all design variations. Hence, actual wind tunnel tests are necessary. The improvements made in computer modeling over the last 20 years should facilitate in the design of electrified leading edges. (See *Product Engineering*, vol. 39, March 11, 1968, pp. 35–6, for a related article.)

18
CHAPTER

Countergravitation

As stated before, electrification appears to not only penetrate inner atomic structures, but also alters the properties of space itself. Scientists have long speculated on the relationships between gravity, space, and electricity, but little progress has been made with finding a common denominator among these properties.

Based on my research, the problem hasn't been clarified for two reasons: Scientists have failed to use visual modeling—specifically, how gravity and electrification act—and researchers have misconceptions about the properties of space itself. Physics textbooks often describe gravity as a force that reaches across empty space and pulls two bodies together (i.e., "action-at-a-distance").

This notion has been wrongly attributed to Newton's views, described in *Principia* (1687). On the contrary, Isaac Newton did not, to his credit, reject the existence of an ethereal medium in space. He wrote, "I have no regard in this place to a medium, if any such there is, that freely pervades the interstices between the parts of the bodies." However, since the character of space was not resolved by Newton or his students, the mystical notion of action-at-a-distance gradually developed. Newton himself labeled this concept "absurd."

The ether, whether thought of as tiny subatomic particles or waves, was considered necessary to account for the propagation of forces. However, by the turn of the twentieth century, when the ether theory was most fully developed, the seemingly paradoxical properties of space still could not be resolved. The result was that abstract mathematics began to replace visual modeling. Even though the ether concepts disappeared by the 1930s, since the 1950s (especially following the work of Paul Dirac) the interest in the properties of space has revived slowly. However, *ether* has been replaced with such terms as *neutrino flux, gravitons* (a quantum unit of gravitation), *soft particles*, *virtual particles*, and *zero-point energy*.

Kinetic gravitational theory

Newton's concept might be considered a static description of gravity. A dynamic or kinetic theory would explain gravity as originating, not in the bodies, but in space itself!

One of the most interesting explanations was that of Georges Louis Le Sage in 1749. His theory might be visualized as follows: Imagine that all space is filled with tiny particles (his term was *ultramundane corpuscles*) traveling in all directions at high speed. Because of their subatomic size, the tiny particles essentially pass through all material bodies. A single body, such as a planet in space, might be pictured as in Fig. 18-1.

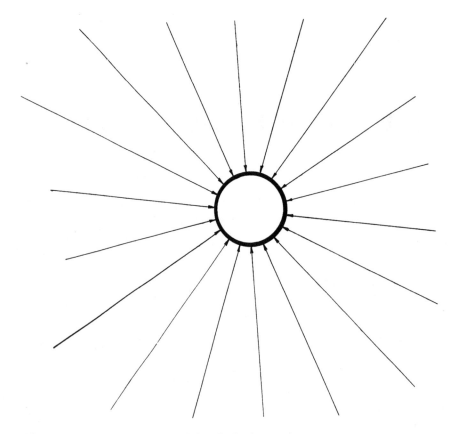

18-1 Action of kinetic space on a singular body.

In the simplified drawing (see also Fig. 18-2), two bodies are pushed together by the superior energy density on the surfaces not facing each other. Each body casts an "energy shadow" on its neighbor so that the ether density is very slightly reduced between the two bodies.

You can now see the importance of visualization as a useful tool in understanding phenomena in physics. These mental models will also help to direct your experimental tests.

You might wonder if there are any natural phenomena or experiments that might shed light on the properties of space. Some optical phenomena, because of the propagation of light, depend on the electrical and magnetic properties of space. Perhaps halos and coronas, sometimes seen around the sun and moon, are indicators. Many anomalous light forms have not been explained.

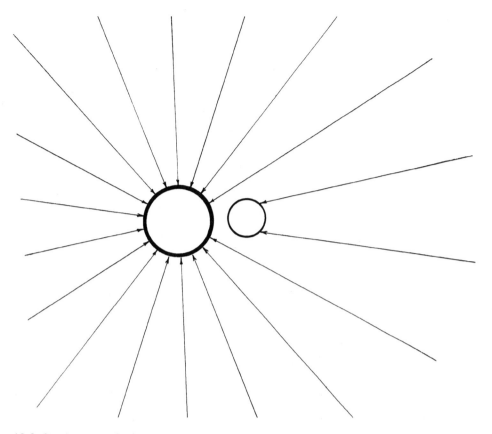

18-2 Gravity—a push phenomenon.

As far as experiments in this subject matter are concerned, American industrialist, Charles F. Brush, published the results of his tests from 1914 to 1929. The results showed that rocks composed of complex silicates of protooxides of nickel and cobalt show a spontaneous rise in temperature in the ambient air during calorimeter tests (1927). In other tests, he found that certain metals and compounds can fall at a slower rate (1 part in 140,000) in a gravitational field! Specifically, bismuth and barium aluminates produced the best results (1924). He attributed these strange results to a slight interaction between atomic structures and gravity waves.

Figures 18-3 and 18-4 give two popular accounts concerning the relationship between electrification and gravitational action.

Dr. Nipher's deflection experiments

This experiment—performed by Dr. Francis Nipher, professor of physics at Washington University, St. Louis, Missouri—is a modification of the Cavendish experiment of 1798. In this earlier experiment, Henry Cavendish used a delicate torsion balance to determine the density of the earth.

The first phase of Dr. Nipher's experiment, performed in 1916 and 1917, is shown in Fig. 18-3. The room had a concrete floor and granite walls, and the equipment

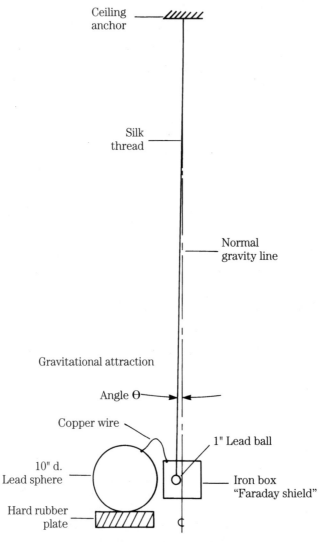

Ceiling anchor

Silk thread

Normal gravity line

Gravitational attraction

Angle Θ

Copper wire

1" Lead ball

10" d. Lead sphere

Iron box "Faraday shield"

Hard rubber plate

Attraction effect of gravity between large and small lead masses (uncharged)

A

18-3 Dr. Nipher's electro-gravity experiment. *Electrical Experimenter,*
March 1918

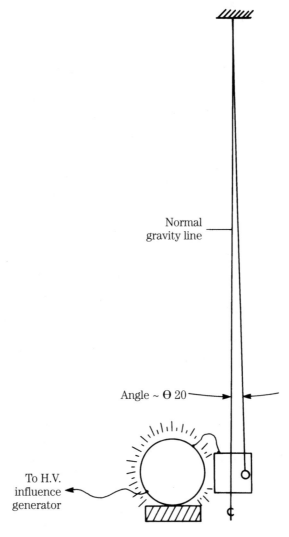

Normal
gravity line

Angle ~ Θ 20

To H.V.
influence
generator

Gravitational "repulsion" between two
masses(large mass charged)

B

18-3 Continued.

was mounted on a massive bench. Thermometers nearby indicated that parts of the apparatus did not vary in temperature from each other by more than 1.5°C. The 1-inch lead ball was suspended with an untwisted silk thread approximately 180 cm long, and centered inside a 5-inch-square iron box or Faraday Shield. A horizontal slit in the box's side, covered with a glass plate, permitted Dr. Nipher to observe scale deflections with a telescope.

Overcoming' Gravitation
By George S. Piggott

For some time past there has been quite a controversy going on regarding the subject of interplanetary communication by means of electric waves. I have been very much interested in the above on account of experiments which I have made and data collected pertaining to gravitation effects on high frequency oscillations and electronic discharges in general. A series of experiments which I conducted during the year 1904, caused me to formulate the theory that interplanetary transmission of electrical impulses was an impossibility on account of the sun's resisting and absorbing influence which vertually isolates our planet from all other electrical vibrations of a lesser tension or power.

Gravitation Suspended in Experiments.

The above theorem was arrived at after I had succeeded in sustaining a metallic object in space by means of a counter-gravitational effect produced thru the action of an electric field upon the above object. A strong electric field was produced by means of a special form of generator and when the metallic object was held within its influence it drew up to approximately a distance of 1 mm. from the center of the field, then was repelled backward toward an earthed contact, going within 10 cm. of the same when it was again attracted toward the field's center but this time getting no nearer than 5 cm. from the polar nucleus. This backward and forward movement continued for some time until the metallic object at last came to a comparatively stable position, about 25 cm. from the field's center where it remained until the power was shut off. While the metallic object was suspended, I was able

to study the effect of the surrounding field and found by means of a powerful microscope, assisted by the insertion of a vacuum tube within the field, that the metallic object (having of course a certain electrical capacity) became fully charged and gave off a part of said charge to and against the surrounding field which tended to hold said object in space, apparently without any other sustaining influence. Around the outside of the metallic object and extending to a distance of about 1/2 cm. was a completely dark belt or space in which there appeared to be no electrical agitation due, possibly, to neutralization caused by the contact of the large incoming energy supply from the field's center with the small oscillating radiations from metallic object. The ever changing action of attraction and repulsion resulted in the overcoming of gravitation. Going farther I will state that the dark belt above mentioned after many tests gave no sign of electrification, a most astonishing phenomenon, inasmuch as its width was but 1/2 cm. In fact, a dark line was shown in the vacuum tube when it was introduced between metallic object and center of field. It is my firm conviction that somewhere on the outer confines of our planet there exists a similar counteracting belt thru which naught but the gravitational vibrations of the sun penetrate, and these vibrations absolutely annihilate or absorb all other less powerful ones.

Therefore, after making many experiments to ascertain as nearly as possible the absolute facts and conditions as they exist, I have come to the conclusion that all electrical disturbances not due to our own radio oscillations, on this globe are due to the sun's electrical activities in semi-inductional contact with our polar extremities.

18-4 Electric levitation of metallic spheres. *Electrical Experimenter,* July 1920

Details of "Defying Gravity."

The illustrations 1 to 4 will possibly give a fair idea of the apparatus used, and the manner in which the experiments were carried on.

Fig. 1 shows general scheme of arrangement of devices. In the lower left hand corner is shown the "ground contact," which can be turned around and placed in any position found necessary, in fact, when metallic object is in suspension, this *ground* can be entirely eliminated.

I have found that any substance within the limits of my experiments can be held in suspension, viz: water globules, metallic objects, and insulators being among those tried. Some materials such as cork and wood exhibit peculiar activities when suspended; a piece of green maple would not rest in one position within the field, but oscillated backward and forward, continuously, going to the field's center, then back to ground.

Heated materials exhibited equally peculiar characteristics: A silver ball 11 mm. in diameter when heated, remained farther away from the field's center than when at normal temperature; upon cooling it gradually drew up to the position it would occupy if unheated.

Fig. 2 shows a generator of the Wimshurst type (improved), the generating or collecting units being entirely enclosed in an insulating case and operated under a pressure of 3 atmospheres; completely dry air *only*, entering case thru drying device attached to air pump shown in Fig. 1. Interior parts of generator will retain quite a powerful charge for a long period of time.

Fig. 3 illustrates suspension stand and field producing electrode, the latter can be revolved in any direction by means of a spring motor shown on upper section of stand.

The small apertures seen in electrode, which is hollow, are there for the purpose of ascertaining the action of the reduced field

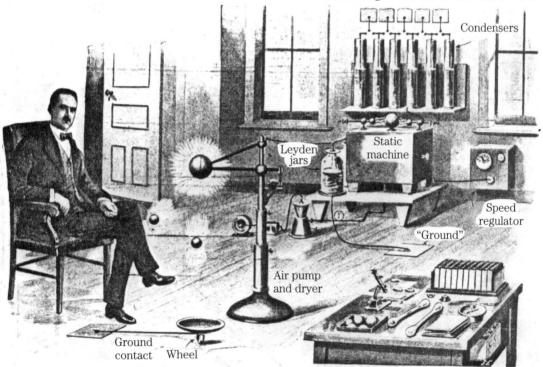

Fig. 1.—This picture shows Mr. George S. Piggott, the author, and his laboratory with the powerful electrical apparatus used, whereby he was enabled to carry on successful experiments in nullifying the effects of gravitation. In other words, he was able to suspend small balls and other objects in the manner shown, the silver balls actually used having weighed 1.3 grams. The diameter of the balls was 11 mm.

18-4 Continued.

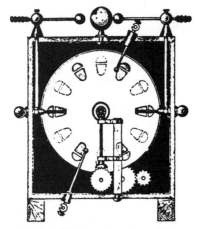

Fig. 2.—Special electro-static machine used by Mr. Piggott in his gravitation nullifying experiments. Which was enclosed in a heavy-airtight compartment, so that it could be operated under several atmospheres of air pressure.

tension at these points, and are also made use of to hold different sized metallic discs, which are cemented to insulating plates, forming condensers, the function of which is to create weak opposite polarities at these points and thus show a reaction on the suspended object and also a greater ocular effect in the vacuum tube.

Fig. 4 is a detailed drawing of the vacuum tube principally used; this is of the spectrum type, without sealed-in electrodes and when introduced into the electric field, glows very brightly at its extremities, especially giving a sharp line bordering the dark space around the metallic object. A very high vacuum is sustained in the tube and it is found necessary to build it of a very perfect insulating glass; the bulb musts be kept absolutely dry on its outer surface.

Different tubes have been used beside the above; corrugated spherical, cone shaped, and cylindrical, with various results.

The electric field produced for suspension experiments is very powerful and

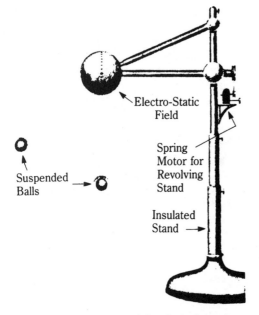

Fig. 3.—A close-up view of the charged metal sphere mounted on a pedestal together with a spring driving motor, whereby the electrode or charged ball could be rotated. The two smaller silver balls are shown as suspended in mid air, the Earth's gravitational pull having been nullified.

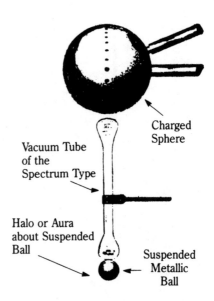

Fig. 4.—Close-up view of vacuum tube of the spectrum type used in studying the aura surrounding the suspended silver balls, while they remained suspended in space.

18-4 Continued.

intense, being detectable with a vacuum tube at a distance of over 6 meters (19.68 feet).

In conjunction with the above and drawing an analogy between the same, I am of the opinion that cometary motion is undoubtedly due to the activity of its compositional elements and their susceptibility to changes of polarity, which, when the comet is far distant from the sun, would be opposite in sign to that of the latter, or when in close proximity to the central orb, would be of the same sign and therefore repelled.

All bodies in process of formation possibly have their cometary stage, and doubtless future experiments will reveal this fact.

Actual Results Achieved by Mr. Piggott

The total power required to operate generator, which was run by electric motor, was about 1/4 K. W. generator; the machine voltage was in the neighborhood of 500,000 when the electrodes were separated beyond sparking distance. The electrostatic charge left on the suspension electrode retained the average object in space for a short length of time, about 1 1/4 seconds after machine ceased rotating.

Some objects such as copper and silver balls, which are of course good electrical conductors, and very nearly homogeneous, when falling toward the earth, after power had been shut off, seemed to slow down when they neared same, and hovered about 2 c.m. above contact for approximately 1 sec. of time before striking same; this was due no doubt to the inductional change of polarity which was imparted to balls almost at the instant of earth contact.

The aura, shown in figure 3, near suspended balls (which in this experiment were made of silver) extended outward to a distance of about 1 c.m. and covered about one-half of the upper hemisphere and a trifle more of the lower hemisphere.

This bluish emanation appeared to be made up of numerous infinitesimal dots or darting particles, each apparently separated from the other by *a very narrow, glowless belt*. Everything was, however, in a constant state of agitation and it was quite impossible to get an absolutely perfect view microscopically, of an individual particle. Different substances have different aura both in length and breadth, and also in luminosity.

The silver balls used in these experiments had an actual gravitational weight of 1 3/10 grams (nearly .05 oz., avoirdupois) and were the heaviest objects suspended at this time, their diameter being 11 mm. as before mentioned in another part of this article.

The largest object suspended was a cork cylinder 10 c.m. long by 4 c.m. diameter (approximately 4 by 1 9/16 inches) which had a copper wire pusht thru its center, and extending beyond its ends to a distance of 3 mm. The weight of above cylinder was 3/4 gram (.002645 oz. avoirdupois).

The behavior of metal spheres used in above experiments was a most interesting spectacle, silver and copper balls floated very steadily on one position and when suspending electrode was revolved, would follow and turn slightly axially, but would not revolve entirely around same, there being a peculiar ''slipping'' effect not entirely accounted for.

18-4 Continued.

Next to this iron box, he placed an insulated 10-inch-diameter lead sphere, with a copper wire keeping this sphere and the metal box at the same potential. To eliminate errors caused by temperature differences, Nipher used cardboard heat shields and kept the observer's body below and away from the apparatus.

Figure 18-3A shows the normal attraction between the uncharged masses. In Fig. 18-3B an influence generator located in the next room is joined to the large mass. After about twenty minutes, the 1-inch lead ball slowly moved to the opposite side with a deflection about twice the normal gravitational attraction regardless of the polarity used.

In the last phase of this experiment in 1917, a torsion balance with two large spheres and two small balls gave the same results. Next, the large lead spheres were replaced with charged metal boxes containing cotton batting. This gave no deflection, eliminating electrostatic force as the cause. Finally, the influence generator was replaced with a low-voltage ac current passing through the two large lead spheres. This also produced a repulsion effect, but one of smaller value.

The full details of the experiment were given in *Transactions of The Academy of Science of St. Louis*, vol. 23, 1916 and 1917. Also see *The Electrical Experimenter*, March 1918, for a related article.

Although Nipher's experiments met a deafening silence when they appeared in the scientific journals, no one came forward with an alternative explanation, even though Nipher was well respected by his colleagues and appreciated for meticulous accuracy in his experiments.

The article from the *Electrical Experimenter* in Fig. 18-4 shows a more popular line of thought.

In Piggott's experiments, the simultaneous appearance of strange luminous halos occurs in conjunction with the effects of levitation. Note that a high-voltage threshhold of about 500,000 volts must be reached before the effect is produced. Of course, by running his Wimshurst in a chamber containing a compressed gas, such as dry air or carbon dioxide, the current output was increased considerably. Oddly, the spheres float. If the phenomenon was simply an electrostatic force, an electrostatic field would first attract, and then repel a metal sphere.

In concluding this discussion of gravity experiments, I mention three other scientists who devoted their lives to kinetic gravitational research:

- Thomas T. Brown, who extended Nipher's work to include the spontaneous self-motion of capacitors (1929)
- Thomas Jefferson See, who developed the mathematical concepts supporting his Wave Theory of Gravity (approximately 1920 to 1950)
- William J. Hooper, who invented two artificial-gravity field generators using the $B \times V$ field (approximately 1968).

A countergravitational force in nature?

In concluding this admittedly speculative chapter, the question arises: Are there examples in the natural world involving measurable gravitational anomalies that can provide inspiration to the experimenter and clues for theoretical consideration? Yes! is the resounding answer. This subject area generally goes by the name *Fortean phenomena* or *Forteana,* named after the prodigious and indefatigable cataloguer of nature's anomalies, Mr. Charles Fort. Several old favorites include:

1. *Niles Weekly Register-Chronicle* (Baltimore), Saturday, November 4, 1815, p. 171.

   Unprecedented phenomenon. We have conversed with several gentlemen, of undoubted veracity, from the country of Ulster, in this state, who all agreed in the following very extraordinary relation: That *they* have conversed with several credible persons from Marbletown, in that country, and they mentioned the names of persons well known to the editor of this paper; and these persons assert, and declare themselves ready to make oath, that the stones lying in two fields there, on several successive days, rose from the ground to the height of three and four feet, and moved along, slowly and horizontally, from thirty to sixty feet; and that a few of them even mounted over the tops of trees! That the persons, who first beheld these astonishing performances, were disbelieved by the neighborhood; but that *all those,* who came to see if there was any truth in the accounts, are prepared to swear to them. The last performance was in an open field without wood or cover near it.

2. *Scientific American,* vol. 43, July 10, 1880, p. 24.

   A Curious Phenomenon. The *Plaindealer,* of East Kent, Ontario, states that a curious and inexplicable phenomenon was witnessed recently by Mr. David Muckle and Mr. W.R. McKay, two citizens of that town. The gentlemen were in a field on a farm of the former, when they heard a sudden loud report, like that of a cannon. They turned just in time to see a cloud of stones flying upward from a spot in the field. Surprised beyond measure, they examined the spot, which was circular and about 16 feet across, but there was no sign of an eruption nor anything to indicate the fall of a heavy body there. The ground was simply swept clean. They are quite certain that it was not caused by a meteorite, an eruption of the earth, or a whirlwind.

3. *The Seattle Times,* Friday, November 24, 1984.

   Cookie cutter? Eerie force uproots big divot. This article by Hill Williams (*Times* science reporter) describes how a massive 3-ton chunk of earth was plucked up and transferred and deposited intact 73 feet away. The mystery site of this event was a remote plateau in north-central Washington state, in Okanogan County, on the Colville Indian Reservation. The displaced earth was discovered by Rick and Pete Timms on October 18 and was believed by them to have occurred after mid-September when they were last at that location harvesting wheat. A small earthquake measuring 3.0 on the Richter scale had been recorded with epicenter 20 miles southwest of the mystery spot on the evening of October 9. However, no tremor was felt at the Timm farmhouse a few miles from the divot. University of Washington geologist Dr. Stephan D. Malone concluded the quake could not have been responsible for such a huge and concentrated upheaval. Another geologist, Bill Utterbach, inspected the hole and ruled out meteorite impact because the hole was not a dished-out crater; it had vertical walls and nearly flat bottom as though it had been cut out with a giant "cookie cutter." But even this was not quite accurate because roots leading to vegetation dangled from the vertical walls of the hole indicated a pulling apart rather than a clean cutting action. The ground around the hole was undisturbed, but pieces of earth dribbled from the huge chunk as it moved; the dribblings traced an arc not a straight line to where the earth mass was found. Greg W. Behrens, geologist with the Bureau of Reclamation at Grand Coulee Dam, also studied the divot, which was an irregular pear shape 10 feet long, 7 feet wide and 1½ to 2 feet thick; it matched the shape and depth of the hole. None of the geologists could offer a plausible explanation for this anomalous but well-studied event. Figure 18-5 sketches the survey of the mystery site.

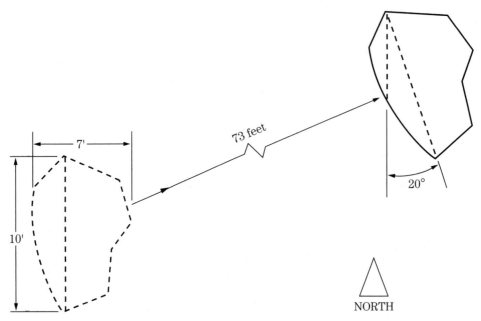

18-5 The "cookie cutter" event.

Many other cases have been recorded down through the years similar to the preceding. They point to the possibility of a suspension of gravity or a countergravitational force that can operate for short periods at specific locations. Perhaps a number of factors need to be just right for its manifestation—moisture level, air-earth electric field potential, geomagnetic field direction and strength, and momentary mechanical vibration at the site come to mind.

This discussion suggests a major paradigm shift in how we view the natural world in which gravity operates, but we must keep in mind that both electric and magnetic forces have the quality of *polarity*; could there also be a negative gravitational force?

Exploding wire experiments[1]

When a wire is suspended just above a white card and electrically exploded, the marks blasted into the card show regularly spaced transverse striations. This phenomenon has been reported throughout history, but never completely understood. Following is a description of the process for setting up such an experiment.

A history of EWP

Exploding wire phenomena (EWP), in which metal wires are caused to explode and dissociate, have a long history. Edward Nairne in 1774 reported his experiments using a glass plate frictional electrostatic generator and Leyden jars connected in parallel to form a high-charge storage battery. He proved that current in all parts of a series circuit is of equal value. Using Leyden jar batteries with a storage capacity up to 100 square feet allowed natural philosophers to study violent, but beautiful, wire explosions reaching 15 feet in length.

Some of the interests surrounding this area of research include:
1. How fuse wires fail
2. Intense light sources for high-speed photography
3. Atomic spectra of the explosion products
4. Plasma state and shock wave production
5. Nuclear-type ionizing radiations from explosions
6. The relation of ball lightning formation and metallic vapors.

When a wire is suspended just above a white card and exploded, the marks blasted into the card show regularly spaced transverse striations. These were first mentioned by Ferdinand Braun in 1905, but their spacing is still not well understood.

---

1 A portion of this chapter was first published in *Electric Spacecraft Journal*, Issue 8, 1992, Leicester, NC.

In his article in *Annalen der Physik*, Braun computed the temperature of a carbon filament discharge at 20,000°C to 30,000°C. He concluded that the confined deposits of metals are not the result of oxidation because the metal vapors are cooled down before the displaced air comes in contact with it again. (There is some oxidation in open-air explosions.)

Following Braun's work little was done until the 1950s. In 1959 William Chace wrote a major book called *Exploding Wires*; his main objective involved studying very high temperatures and the plasma state of dissociation.

In the relation $E = \frac{1}{2} CV^2$, which expresses the energy stored in a condenser storage system, we see that the energy, and therefore the temperature available for vaporizing metals, increases directly with the capacitance C and with the square of the applied voltage, V. Figure. 19-1 illustrates what this energy can accomplish.

19-1 Appearance of the card after sending the discharge through silver wire $\frac{1}{300}$ of an inch.

Exploding wire experiment

Because exploding wire experiments can give insight into electric discharges like ball lightning (see A. T. Jone's experiment Fig. 20-9), I describe an inexpensive circuit and experimental setup for testing short pieces of wire.

Warning

Because of the danger involved, this research is not for inexperienced electricians. Only a seasoned experimenter having a clear head is qualified in this area, since charged capacitors show no mercy—they can injure or cause death in an instant.

- Always wear ear and eye protectors, and if needed, build a sound-absorbing chamber when the noise creates a disturbance.

- Older capacitors may contain PCB oil. Write to the manufacturer for details.
- The use of gas or electric welding filters with eye protectors is recommended.
- Never wear contact lenses when looking at powerful electrical discharges. Intense discharges can "weld" contact lenses to the cornea and cause permanent blindness.

Figure 19-2 shows the circuit diagram. This is a simple half-wave transformer rectifier; the high-voltage transformer may be a furnace oil-burner transformer or a neon sign transformer with a minimum current of 1 milliamp and at least 5000 volts (ac) output. One terminal is grounded and also joined to the discharge capacitor and to the grounding clip on one side of the sliding spark gap. The capacitor may be a laser discharge capacitor or filter capacitor rated at 0.4 microfarads, 10,000 Vdc to 90 microfarads at 1500 Vdc. The other terminal of the transformer joins to the filament tube circuit of a #1B3GT high-voltage diode tube (available from Antique Electronic Supply, Tempe, Arizona).

I place this tube, tube socket, and battery holder on a well-insulated acrylic base. The top terminal of the tube is connected to a variable resistance with a range of 10 to 100 meg (million) ohms and 100 watts capacity. This resistance helps charge the capacitor slowly so that it is less likely to be ruptured by the high current.

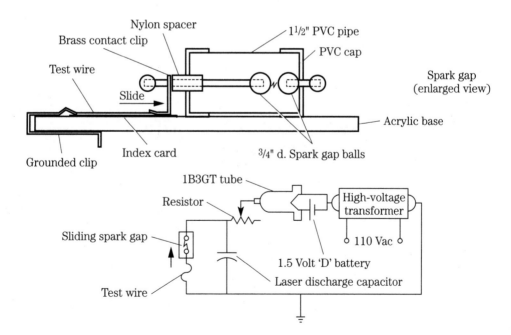

19-2 Circuit for exploding wires.

I had to make this resistor up using an alligator clip that slides along a 10-inch-long section of bamboo skewer that has been coated with graphite powder to make it conductive. This will smoke a bit when power is applied. A terminal lug clamped onto the other end of this variable resistor connects with the high-voltage terminal of the capacitor and to the fixed high-voltage side of the spark gap.

In use, the test wire to be exploded, which should not exceed 0.006 inches diameter × 5 inches long, is taped with aluminum foil tape to a white, stiff 3-×-5-inch card, such as an index card or poster board. The foil tabs are slipped under the two brass clips shown at the left side of the spark gap. This is on the *grounded* side of the circuit for safety reasons. The ¾-inch spark gap ball on the left is separated slightly from its neighbor; power is supplied for about 10 seconds and then shut off. I now close the gap using an insulated rod or discharge tongs so that the charged capacitor is instantly drained through the wire to ground.

You might need to enclose the entire circuit in a Styrofoam box when the explosion disturbs your relatives (now distant?) or pets. Since the entire output of the transformer is stored over a time interval of up to 1 minute, then short circuited in a few millionths of a second, this work is not for the faint-hearted! The flash, brighter than the sun, sounds like a shotgun blast but leaves some beautiful traces on the cards. Samples using the above circuit are shown in Figs. 19-3 through 19-7.

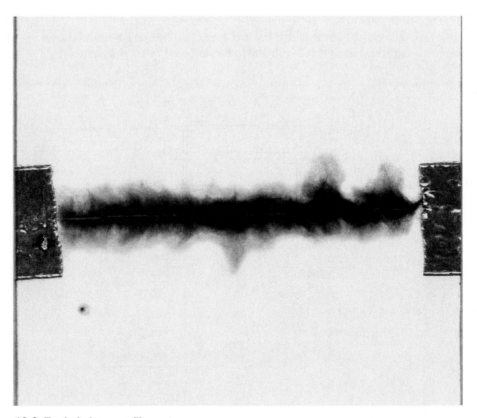

19-3 Exploded copper filament.

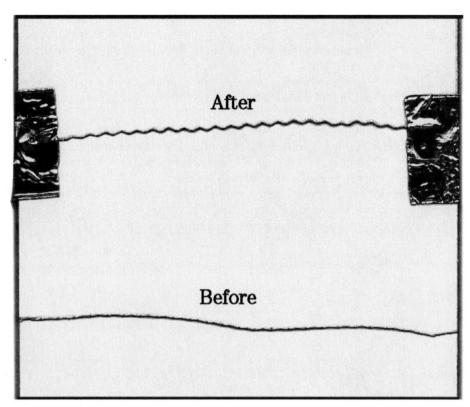

19-4 Aluminum screen wire contraction after discharge.

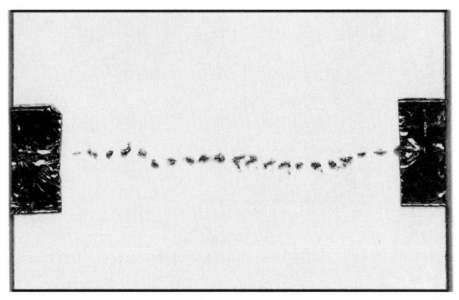

19-5 Aluminum wire reduced to evenly spaced drops.

Figure 19-3 shows a copper filament, 0.005 × 3 inches long. The dark rectangles at the ends are foil tape. The color is grey, black, and yellow, with a small dot below where a red hot copper ball touched. Figure 19-4 shows aluminum screen wire before and after a medium discharge. The corrugations are deeper because the wire has contracted. Figure 19-5 shows how aluminum wire breaks up into drops evenly spaced. Figure 19-6 is a 0.002 × 5-inch filament of #3 coarse steel wool with colors grey and brown.

Figure 19-7 shows a copper filament given an oxbow shape with a diameter of ⅜ inch and a ¹⁄₁₆-inch gap below. Following discharge, the current arcs across the gap rather than going around the bow. This is because electricity has inertia and the effect was named by Mr. Faraday the "lateral discharge."

When finished with an experiment, use only heavy-duty discharge tongs, always connecting with the grounded terminal first. I never use a switch in the power supply—just a cable plug—because switches can fail.

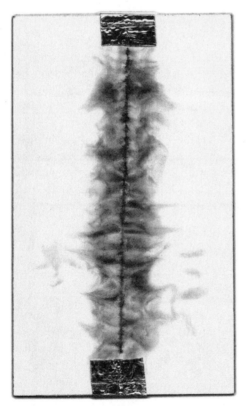

19-6 Steel wool filament, 0.002 inch × 5 inches long.

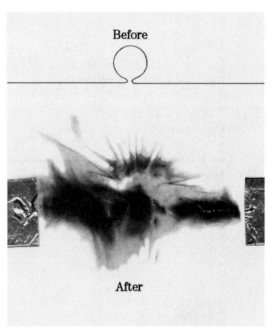

19-7 Copper "oxbow" filament with short gap.

20
CHAPTER

Unusual
electric discharges

This chapter features a few of the strange anomalies in electrical science that will be of considerable interest to those making electrical investigations.

Lightning shadowgraphs

Lightning shadowgraphs might be called nature's method of photography. These shadowgraphs are of shadow-pictures of objects projected on nearby surfaces from bright flashes of lightning. For example, the *English Mechanic* (1892) reported an incident in Errol, England. A telephone repairman, while fixing a fused telephone wire damaged by lightning, found an image of the roof of a nearby house on one of the porcelain insulators. Apparently, the brilliant light and vaporization had flashed the image onto the smooth surface of the insulator.

The second example (Fig. 20-1) comes from *Scientific American Supplement* (1904).

Occasionally, these images have been fairly permanent. The image of a lady's face at her bedroom window watching a thunderstorm was said to be flashed onto the window pane. After many years, the impression gradually faded away, perhaps by erosion of the glass surface. The chemical condition of the nearby surface is important.

Tornadoes as electrical machines

Several years ago, I experimented with the effects of high-voltage direct-current discharges onto moist semiconducting surfaces. The substances—including granite, marble, agate, limestone, sandstone, white chalk, plaster of paris, slate, and unglazed clay—were chosen for their fine porosity and ability to absorb moisture.

A most unusual discharge presented itself while I was working with unglazed clay flowerpots. Figures 20-2 and 20-3 illustrate the setup for producing miniature

Remarkable and rare effects of lightning.

An excerpt from the Annals of the German Hydrographic Bureau furnishes us with a bit of information at once interesting and astonishing in its effects. While on a voyage recently from Hamburg to St. Thomas the second officer of the Hamburg-American liner "Galicia," being on the bridge during a terrific electrical display, observed the following phenomena, which he carefully noted, and which it is our privilege to present to our readers. In advance it may be remarked that all the wood and iron work about the bridge had been painted gray. In changing his position he casually removed his hand from a cabinet on the bridge immediately after a particularly brilliant flash of lightning, and what was his astonishment to notice an exact counterpoint of it in silhouette upon the cabinet, and to add to his amazement the picture remained imprinted fully five minutes. Such a spectacle was well calculated to incite the officer to further observations, which he carried out with like results. Among others he placed an observation instrument upon the cabinet, and waiting his opportunity removed it just after a vivid flash, to find the shadowgraph perfect in detail, even to the cross-hairs over the objective plainly visible upon the surface.

Since the ship's deck was also painted gray, he determined to try a further experiment; and with this in view threw down upon the deck an annular cork life preserver, allowing it to remain untouched for several successive flashes.

In throwing it down, whether with intent or otherwise, the ship's name and hailing port, "Galicia, Hamburg," painted upon the cork ring, fell downward next the deck. When the ring was removed, the shadowgraph was plainly seen, and what was more, the inverted letters in more somber tones could be distinctly read. Until it had entirely disappeared, the watch tolled off seven minutes, the additional duration resulting from the effect of the several consecutive flashes. Keenly awakened by a spirit of investigation, the officer experimented upon the galvanized ironwork sustaining the bridge, which was, as before said, also painted gray.

From this he failed to elicit any response, while all the woodwork seemed particularly sensitive. Moreover, it was discovered that success depended upon the wet or moist condition of the painted surfaces; upon dry objects of the same color no pictures were obtained.

In discussing the phenomenon, the annals remark that should an attempt be made to explain the pictures by declaring that the lightning of itself had nothing whatever to do with their appearance, but rather that the different objects placed upon the cabinet and the deck had absorbed the moisture, and caused a dry spot surrounded by a wetted surface, making it distinguishable from its surrounding by a shade of color, it would hardly explain the presence of the cross-hairs over the objective, which are contained entirely within the instrument. Nor would such a hypothesis hold when compared with the experiments upon the ironwork.

A more plausible elucidation of the occurrence would derive from a chemical examination of the constituents of the paint used, which might disclose some phosphorescent properties of the ingredients. Upon request the Hamburg-American Line furnished the German Marine Observatory with some of the liquid, which by some inadvertence or carelessness has been lost before it could be used. Having aroused an interest in the proper accounting for such amazing displays, the government desires that observations be continued, and in cases of recurrences, either some of the paint or some object covered with it, which has given the abnormal results, be sent to the Lighthouse Board for official investigation.

20-1 Remarkable and rare effects of lightning. *Scientific American Supplement, 1904*

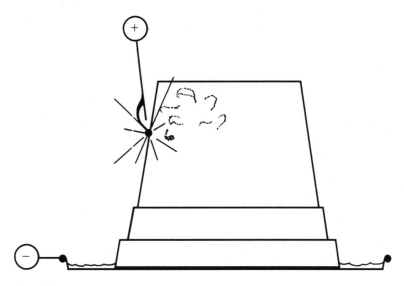

20-2 Miniature electric tornado.

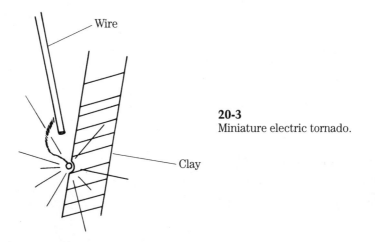

20-3
Miniature electric tornado.

electric tornadoes with a white fireball tip. A 1-inch-long section of plain steel piano wire, 0.015 inch in diameter, is positioned $\frac{1}{16}$ inch from the clay surface. The pot sits in a shallow metal pan with water covering the bottom. The power supply is a full-wave transformer rectifier that has an output of 10,000 Vdc at about 0.7 to 1.0 milliamp (7 to 10 watts). The wire is made positive, and the pan is negative.

The unglazed clay's properties are crucial. At the Ceramics Research Laboratory, University of Illinois, analysis comparing clay pots indicates that the acceptable clay, on which the vortex forms, has a dry surface resistance of infinity and, when dipped in water and the excess wiped off, a surface resistance of 300,000 ohms. Place the probes of the ohmmeter 1 cm apart for this indication. The clay color is light red; that means it contains a smaller percentage of iron oxide. (Thanks to Dr. Relva Buchanan for determining the characteristics of the clay samples!)

Moisten the selected flowerpot with your finger dipped in water and place it as shown. Turn on the power and the discharges will remove moisture, increasing the clay's resistance. In the dark, when the fireball forms, the discharge either squeaks or is silent. With the positive wire near the clay, a dazzling pure white ball, about 1 mm in diameter, will form on the pot in an electric tornado-vortex, varying between ⅛ and ⅜ inch long. The ball will slowly traverse the surface in a sinuous movement, seeking out a path of preferred resistance.

When I examined the ball through a #4 gas welder's filter, rays from the fireball were still visible. The amazing thing to me is that the heat from the tiny fireball was so great that it permanently etched a black path into the clay. Figures 20-2 and 20-4 show the characteristic track signatures. When the polarity is reversed, the tip of the steel wire (−) often glows white hot, and the clay remains cool.

20-4
Fireball-tipped tornado on moist clay. Film: ASA 400, F/5.6, ⅟₁₅ second, scale in millimeters. Note second fireball below.

20-5 Closeup photo. Note tornado-like vortex. Cathode is slate tile. Power was 16,800 Vdc at 30 milliamps.

How does this unusual discharge relate to real tornadoes and waterspouts? Many good descriptions of tornado lights with fireballs and internal lightning bolts exist, but I found only one case with a fireball maintained at the terminal end (as in our experimental condition). One rare account in Fig. 20-6 is from the British journal, Weather (1949).

UNUSUAL DAMAGE BY A TORNADO
By E. S. DAVY
Assistant Director, Royal Alfred Observatory, Mauritius

The island of Mauritius in the South Indian Ocean is in a region in which tropical depressions or ''cyclones'' frequently develop and move; small tornadoes or whirlwinds are, however, rarely reported here.

On May 24, 1948, at Curepipe, Mauritius (approximately 1,850 feet above sea level), a phenomenon of the tornado type exhibited a very interesting feature which the writer has not seen mentioned in reports of other tornadoes. During most of the day the weather was mainly fair at Curepipe and convection appeared to be no more intensive than is normal for this region. In the middle of the afternoon a whirlwind touched down to earth over a hard tennis court, causing considerable damage to its hard compact surface. A trench running in a north-south direction, 60 feet long and 1 to $2^1/2$ feet wide, was cut in the bare surface of the court to a depth varying from 1 to 4 inches. The material lifted from the trench was all thrown to the west to a distance of 50 feet; pieces weighing about one pound were thrown as far as 30 feet. The surface material was slightly blackened as if by heating, and a crackling like that of a sugar-cane fire was heard for two or three minutes. The court was made of a ferruginous clay, which packs down to a surface more smooth than that of the hard tennis courts usually made in Great Britain.

Unfortunately, there were no very reliable witnesses of the phenomenon. The impressions gained by two servants of the tennis club who saw the incident differed considerably on important details. One claims to have seen a ball of fire about two feet in diameter which crossed from a football pitch to the tennis court through a wire-netting fence without leaving any evidence of its passage until it bounced along the court, making the trench in the surface before disappearing completely.

The high winds and ascending currents in this small tornado were unusually irregular in their action; on the tennis court an umpire's chair weighing about 50 pounds was picked up from 50 yards to the west of the trench, carried upwards to an estimated height of 60 feet, said to have been broken whilst in the air, and dropped in pieces about 20 feet to the east side of the trench, that is, on the side opposite to the pieces of court surface. A hundred yards to the south of the tennis court the roof of a small building was lifted off and dropped nearby. About two miles to the north a rather insecure building was blown over as the whirlwind moved in a northerly direction. No other evidence of damage to buildings or vegetation was reported nor could any be seen from a hill-top which was in the track of the phenomenon, between the damaged sites of which a good view was obtained.

20-6 Unusual damage by a tornado. *Weather,* 1949. By permission of the Royal Meteorological Society

Caution to experimenters using high-wattage power supplies: The evolution of heat might be so great as to cause flash-steam explosions at the clay's surface. Some shielding for your eyes should be included in your setup. Also, be sure that the wire electrode does not vibrate; this action destroys the vortex formation.

If you are interested in tornadoes, their formation, and prediction, consider the following avenues and questions:

- Does a relationship exist between tornado "alleys," and ore deposits and underground streams?
- Based on our experiments, electric polarity is important. Does the anvil cloud, therefore, have a high-positive charge where the vortex forms and a negative-charge concentration on the ground below the cloud?
- Would studying the miniature electric tornadoes with high-speed infrared and ultraviolet photographic methods, and the Doppler shift (for determining the rate of fireball rotation) help in understanding the mechanics of vortices?
- Is there a connection between trailer parks and tornado formations? Trailer roofs present a large aluminum surface to the sky; this might strengthen the air-earth electric field.

Indispensable to any study of tornado anomalies is the excellent source book *Tornadoes, Dark Days, Anomalous Precipitation and Related Weather Phenomena* (1983) by William R. Corliss.

Electrical signature of tornadoes

I wish to point out a factor that needs to be studied by scientists who gather data by chasing tornadoes. Many videos show a close relationship between tornadoes and intense lightning. Study of the *air-earth electric potential gradient* while powerful storms develop may yield an electrical signature indicating tornado development. One experimenter, traveling with an oscilloscope (which measures electrical potential) in his car, passed into a storm with nearby tornado activity. With voltage set at 50 millivolts per division, first, a perfect-looking ±100-millivolt sine wave appeared on the scope screen, making a single cycle lasting about 1 second. The beam trace became smooth again. In a few minutes, the beam trace started making irregular continuous waves of ±50- to 75-millivolt amplitudes across the screen. This lasted 10 seconds and then stopped. Several minutes later, a second single-cycle sine wave appeared like the first one. During this strong storm, the beam trace was about 25 millivolts negative. The tornado in the area may have given the oscilloscope screen this slow undulating motion as a specific electrical signature. The scope used in this mobile experiment was a digital Philips PM 3350 50-MHz band width set on a 0.5-second per division slow seep; its high voltage lead was connected to an outside antenna. Any additional factors that can be added to the computer modeling of tornado formation will increase warning time, where extra minutes are crucial.

The electrical entities (fireballs)

The various names for self-sustaining electrical forms have included electric meteors, bolides, fireballs, and lightning balls. I prefer the term entities because they are self-sustaining embodiments of electrical force.

In its ordinary form, an electrical entity appears as a smooth sphere emitting light in often beautiful colors. However, its bewildering variations include spheres that are smoky black, globes with spikes or diverging rays, double and triple balls connected by luminous threads, globes with long tails, rod-shaped lightning bolts, and luminous rings.

The "bible" of eyewitness accounts is, at present, the 170-page document Der Kugelblitz by Walther Brand (1923), NASA translation (1971). Please see the bibliography for other sources, since Brand's work mainly covers the classic ball shape.

Drawing from several sources, which include both naturally formed and artificially formed electric entities, I have sorted out several conditions that are most often present in the formation stages. The list includes:

- The presence of aerosols, such as a fine mist. The air is usually at the saturation point. Also, the presence of excessive dust, soot, or finely divided metals and semiconducting stone particles is frequently noted.
- The presence of hygroscopic porous or fibrous bodies including adobe, stone, wrought iron, metal oxide on conducting wires, plaster and chalk, talc, gypsum, or clay.
- Discontinuities in metal conductors, including rapid changes in cross section, sharp bends, projecting points, and in the case of chain conductors, poor contact between successive links.
- A very slowly rising electric-field intensity that reaches a high potential without causing an actual breakdown. It evidently takes some time for the air molecules surrounding the site to become mobilized to an electrified state. Suspended drops of moisture might help this process to occur.
- Occasionally, the initial entity formation requires a shock in the form of a thunderclap or falling object. Perhaps sound waves disturb the air molecules, which precipitates the entity's appearance.

In view of these requirements, it becomes clear why this phenomenon is seldom seen in the conventional experimental laboratories. Philosophically, scientists prefer clinically clean and highly uniform conditions; "dirty" experiments that introduce too many variables are avoided. However, the natural environmental conditions should be present if anomalies are to appear.

I need to expand on the fourth condition listed above. It is important to clarify what type of motion found in nature could cause a ball lightning to form and maintain itself. Spinning motion (like a top) would be one form. There are many eyewitness accounts that describe spinning fireballs. This could explain the erratic movement of the ball just as a spinning hurricane's path is very hard to predict.

Another mode of cyclical motion that could form a ball lightning is the spherical or coreless vortex ring. (Spherical vortex rings of smoke were first described by the Irish scientist Robert S. Ball in 1868. This vortex ring has a very small hole and so appears to be a ball in shape, like an onion head.)

Experiments with smoke rings show they have a quality that is shared by ball lightning, which is the most fundamental property for both: **localized persistence of individuality**. This means that once initiated, smoke rings and ball lightning maintain themselves and are not easily dissolved. In 1965, I decided as an experiment to shoot an uncharged smoke ring past the charged terminal of my 500,000-volt Van de Graaff generator. I was convinced that the charged terminal would bend its path. I was quite surprised to see the electrically neutral smoke ring pass within 2 inches of the terminal yet continue on in a straight line. Very many eyewitness accounts state that ball lightning is unaffected by grounded conductors or power lines, so this could be accounted for by the ring vortex form of internal motion, which maintains its own integrity and stability.

It would be instructive to consider an eyewitness account in which ball lightnings form repeatedly at the same spot over a short time period. These events are very rare, but I reprint an old favorite here:[1]

> After a drought of several months in Grenoble, the rain finally started to fall on Wednesday afternoon, October 2, 1895. Even though there was no lightning or thunder, the weather was still oppressive and sultry. The rain fell under the same conditions during the whole day. At about 8 o'clock, when I walked up to a window, I suddenly saw a large fireball appear on the point of an iron rod, which was mounted on the roof of a neighboring house in order to support telegraph wires. Since the distance between me and the house in question was only 100 m, I could observe the phenomenon very clearly. This ball, whose contour, despite the luminous emission, seemed to be sharply outlined, may have been 30 cm in diameter. It has the lustre and appearance of a powerful electrical furnace. From the point of the iron rod there emanated a continuous stream of fairly large sparks, which seemed to originate from the white-hot iron platelet. These sparks were remarkable similar to those produced by a blow of a piledriver. The stream of sparks was directed downward.
>
> After a while, which I estimate to be about 40 or 50 seconds, the fireball suddenly split into three smaller balls, each the size of a children's balloon like that sold on the street. The sparks stopped immediately and the three balls, which were identical in appearance to the first fireball, seemed to roll down the roof as if merely under the influence of gravity. Upon arriving at the gutter (or perhaps upon touching it, since at this instant several sparks appeared again) all three balls disappeared without producing a detonation.
>
> Almost immediately thereafter a second fireball appeared on the point of the same rod. However, it vanished after 2 or 3 seconds without a sound. At the same time a stream of sparks emanated from the point of the rod and was similar in size and color to the previous one. This stream followed the same direction as the first one.
>
> I consider it relevant to add that at about 5:30 a trustworthy person had seen two fireballs at the same place and with the same time interval of several minutes, and had exchanged a few words with me about this.
>
> The next morning I noticed that the iron rod was no longer upright, but exhibited a very conspicuous kink.

[1] Comptes Rendus, vol. 121, 1895, page 596

20-7
Vortex ring modeled with toy
"Slinky."

I subsequently discovered that on the same evening several persons had observed similar phenomena at different places in the city.

The pointed iron rod in the above observation could provide the radial symmetry for a ring vortex's initial formation. Liquid vortex rings falling under gravity's influence can split up into several smaller rings. This splitting effect would be difficult to explain if a ball's internal motion was not a vortex. Figure 20-7 models the vortex ring using a toy "Slinky." This vortex theory should be considered as one of several possible modes by which fireballs appear.

Historic entity experiments

I now provide three specific and quite fascinating cases where these self-contained entities have been produced artificially. The equipment required for these experiments is neither exotic nor expensive.

The first of these experiments appeared in the Electrical Experimenter (Fig. 20-8). In Weisiger's and Leduc's experiments, the sharp-pointed electrodes and the semi-conducting surface of the film, mentioned in item #3 of the requirements, are notable factors.

Another example of these experiments is from the journal *Science*, n.s., 1910 (Fig. 20-9): A.T. Jones' report, a very simple short-circuit experiment, is at least partly covered by item 3. When duplicating experiments using short circuits, the condition of the wire—if it is cut or bent near the contact point—will alter the results. Unfortunately, there is no mention of moisture or oxides here, just of the tiny 1-mm copper ball left behind. The color of the ball should be considered indefinite, since Jones expressed some doubt.

My last and most interesting report, every bit as important as Franklin's kite experiment, is taken from two rare sources: *De L'electricite des Meteores* (vol. 1) by Professor Abbe Bertholon (1787) and *A Complete Treatise on Electricity in Theory and Practice; with Original Experiments* (vol. II) by Tiberius Cavallo (1795).

I retyped the second account because of the poor quality of the paper and the hard-to-read script common to eighteenth-century writings; the sketch (Fig. 20-10) is found in Bertholon's description.

"Ball Lightning" Experiments
By Samuel S. Weisiger, Jr.

In the January, 1916, issue of the *electrical experimenter* you publisht a discourse on "Ball Lightning," and gave instructions for the experimental production of it. Thru the kindness of Mr. Porter, Instructor in Physics at the Allegheny High School, I have been able to make the several photos accompanying this letter. Under each photo there is given a short description of the circumstances under which each discharge was made and the phenomena connected therewith.

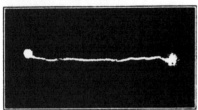

This is Another "Freak" Discharge. The Ball Travelled in a Very Crooked Path to the Positive Electrode, and Here Exploded. The Force of the Explosion Was So Great that a Part of the Spark-Ball Was Thrown to the Other Side of the Positive Electrode, from Whence It Continued to the Positive Electrode.

In making these photos a 75,000 volt Toepler-Holtz static machine was used. The distance between the sharp metal points was from 5.5 to 6 centimeters. This distance must be found by experiment, and altho it is absolutely essential to have the correct distance between points, it will nevertheless differ with the capacity of the static machine.

Much trouble will be encountered if the sharp points, used to produce the discharge are not free from grease and highly polished. The best way to polish the points is to take a little powdered chalk (blackboard chalk which has been scraped to a fine powder with a knife) and put it on some kind of cloth and turn the point of the electrode, at the same time giving considerable pressure to the cloth where the point is being turned.

The best connection for the electrodes was found to be obtained by means of two brass chains.

Two large-sized sharply pointed darning needles suitably mounted form admirable electrodes. It is practically impossible to use blunt needles.

There will be much trouble in finding the correct spacing for the electrodes and it will probably require some experimentation. In any case the spacing is dependent on the power of the static machine.

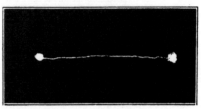

Some Trouble Was Encountered in Getting this Spark-Ball to Form. Evidence of This is Shown by the Plate Being Exposed By a Tiny Charge Or Burst of Light On One Side of the Negative Electrode. The Uneven Course of the Spark-Ball is Clearly Defined.

When the plate is put under the electrodes be sure to get the emulsion side up as the discharge occurs better when the plate is placed in this manner.

When the plate is under the electrodes and the static machine has been started, the spark ball should form very quickly. After the ball has detached itself from the electrode, turn the machine very slowly in order to expose the plate longer. The rate of travel of the spark ball is proportional to the speed of the static machine.

Should the machine be stopt before the spark ball reaches the other electrode, the plate will only show the path of the ball to that point.

Knowing that there is considerable interest in these "Ball Lightning" experiments we have republisht below the original directions for producing ball lightning in the laboratory as outlined by the famous French scientist—M. Stephane Leduc. His experiment makes possible the production of a slowly moving globular spark not easily obtainable in any other way, in so far as we know.

20-8 "Ball lightning" experiments. *Electrical Experimenter, 1919*

To produce this imitation ball lightning it is necessary to employ two very fine highly polished metallic points, each of which is in connection with the positive and negative poles, respectively, of a static machine of small or medium size. These two metallic points must rest perpendicularly, as our illustration indicates, on the sensitive face of a gelatin bromid of silver photographic plate, which is placed on a metallic leaf, such as tinfoil. The two metal points are spaced about five to ten centimeters apart. When the static machine is operated an effluvium is produced around the positive point, while at the negative point there is formed a luminous *fireball* or *globule*.

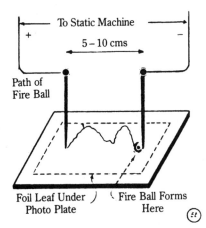

Scheme for Producing Ball Lightning in the Laboratory with Static Machine. Photograph Plate and Two Needles.

Now, when this globule has reached a sufficient size, it will be seen to detach itself from the metallic point, which then ceases to be luminous, and the globule will begin to move forward slowly over the surface of the plate, taking various curved paths and eventually it will set off in a direction toward the positive metal point. When it reaches this electrode the effluvium is extinguished and all luminous phenomena ceases. Further, the static machine acts as if its two poles were short-circuited, or, in other words, united by a conductor.

The velocity acquired by the luminous globule as it travels is quite slight, it taking from one to four minutes for it to traverse a path of six centimeters in some cases, and before reaching the positive electrode the globe bursts into two or more luminous balls which individually continue their journey to the positive electrode. On developing the photographic plate (which, of course, should be placed under a ruby light while the foregoing experiment is conducted) there will be found a trace on it of the exact route followed by the spark globule—the point of explosion, the routes resulting from the division, and the effluvium around the positive electrode point.

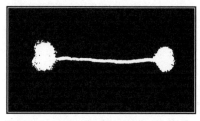

This is Probably the Best Photo of the Set, the Spark-Ball Being the Largest Obtained. You Will Notice the Manner in Which the Ball Broke Into Two Parts and Each Part Proceeded to the Pole. The Effluvium Around the Positive Pole Shows Signs of a Violent Explosion As Will Be Noted by Closely Examining the Tree Formation Made By the Bursting Spark-Ball.

Also, if one should stop the experiment before the globule's arrival at the positive electrode, the photograph will only give the route to that point. The fireball takes for its course the conductor, which apparently short-circuits the static machine. If sulfur or some other powder is thrown on the photographic plate while the experiment is being conducted, and also while the ball is moving, its path will be marked by a line of aigrettes, looking very much like a luminous rosary.

[*The Editors will be glad to hear from any of our readers who have made experiments in this direction. Photographs are particularly welcome.*—Ed.]

20-8 Continued.

A LABORATORY ILLUSTRATION OF BALL LIGHTNING

In Dr. Elihu Thomson's address at the opening of the Palmer Physical Laboratory at Princeton University he made, with regard to ball lightning, the statement, ''The difficulty here is that it is too accidental and rare for consistent study, and we have not as yet any laboratory phenomena which resemble it closely.'' This suggested to me that a phenomenon which I witnessed some six or seven years ago might be worth recording.

With a copper wire a student accidentally short-circuited the terminals of an ordinary 110-volt circuit. I happened at the time to be a few meters from him and to be looking toward the terminals. At the instant of the short circuit I saw an incandescent ball which appeared to roll rather slowly from the terminals across the laboratory table and then disappeared. As I remember it, I should say that the ball may have appeared to be about three centimeters in diameter. I think no one else in the room saw anything more than a flash of light—much as if a fuse had blown. On the table where the ball had rolled we found a line of scorched spots, as if the ball had bounced along the table and had scorched the wood wherever it touched. As I remember them, these scorched spots were rather close together, perhaps not more than one or two centimeters apart. In the top of the table was a crack perhaps a millimeter or two wide, and at this crack the scorched line ended. In a drawer immediately under this crack we found a tiny copper ball, perhaps a millimeter in diameter. Apparently the ball that rolled along the table was incandescent copper vapor, although my memory of it is rather of a yellow-white than of a greenish light.

The above suggested the possibility of a laboratory study of a phenomenon which may very possibly be similar to that of ball lightning, but I have never attempted to repeat the experiment.

A. T. Jones
Purdue University

20-9 A laboratory illustration of ball lightning. *Science, n.s., 1910*

Extract of a Letter from Mr. Arden, Lecturer, in Natural Philosophy, Dated September 25, 1772

About fourteen or fifteen years ago, in the presence of Wm. Constable, Esq; at his seat at Burton Constable, in Holderness, I made the following experiments:

I placed a large coated jar, that would hold three or four gallons, directly under the prime Conductor of a very good electrical machine. The prime Conductor was at least eight or ten inches above the top of the jar, and the communication was made by a brass wire, bent at one end over the prime Conductor, and the other end passed through a small glass tube (contrived by Mr. Constable to prevent the electric matter from easily flying off) was suspended in the middle of the jar, and had a small piece of brass chain fastened to it, that rested on the bottom of the jar.

I then began to turn the wheel, and, after turning about 100 or 150 times, as low in the jar as I could see for the coating, I perceived a ball of fire, much resembling a red-hot iron bullet, and full three quarters of an inch in diameter, turning round upon its axis, and ascending up the glass tube that contained the brass wire, which was the Conductor to the inside of the jar.

I immediately asked Mr. Constable, if he saw the ball of fire? He said, Certainly. I said, I will turn on. He answered, By all means. I kept turning the wheel, and the ball of fire continued turning upon its axis, and ascending up the glass tube till it got quite upon the top of the prime Conductor. There it turned upon its axis some little time, and then gradually descended, turning upon its axis as it had done in its ascent, and so continued till it was so much below the top of the coating that we could no longer see it. But soon after

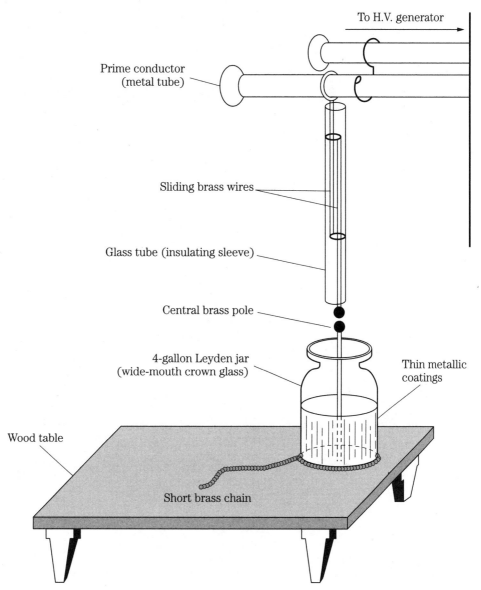

20-10 The Arden and Constable experiment. A Complete Treatise on Electricity in Theory and Practice with Original Experiments, vol. II, Tiberius

this, a very great flash was seen; a large explosion was heard, and a strong smell of sulfur was perceived all over the room; a round aperture was cut through the side of the jar, as fine as if it had been cut with a diamond, rather more than three quarters of an inch in diameter, and between two and three inches below the top of the coating, and the coating was torn off all round the aperture, about three or four inches in diameter. The jar was a pretty strong one, of crown glass.

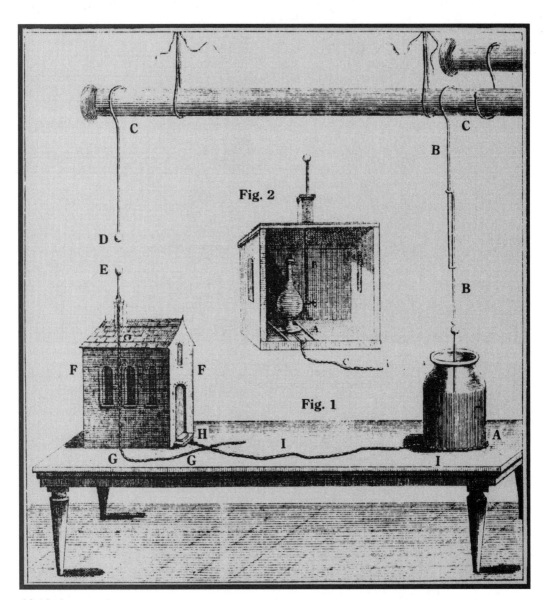

20-10 Continued.

I then took another jar, so like the first, that when both were whole I could not easily perceive any difference between them. I then attempted to charge this jar, in the same manner as the other, and we both observed it very accurately. No ball of fire was seen, but presently the jar discharged itself with a great flash and explosion, and at about the same part as of the first jar; but instead of the aperture which was made in the first jar, there was a circle about three quarters of an inch diameter, as white as chalk, and the coating torn off round about it as before. Upon touching the white part, it dropped out, and appeared to be glass in a fine powder.

We broke several other different-sized jars that day, (which made Mr. Constable say we were in great luck) but without any thing else remarkable.

The first experiment was made soon in the afternoon of a clear day, and the machine stood directly between us and a window, which was not above a yard from it. I don't hear that this ball of fire has been produced by art by any one else, to this day, although it is often produced by nature.

I had the pleasure of seeing Mr. Constable this day, and of reading the account of these experiments to him, and, to the best of his memory, he thought the whole was strictly true.

Mr. Constable thinks it would not be difficult to repeat the experiment, and to produce the ball of fire at any time, provided the jar is large, and not coated too near the top, and that the wire communicating from the prime Conductor to the inside of the jar is made to pass through a small glass tube (which is certainly of great advantage in making experiments of this kind) and that the machine acts very strong. If not, it will be in vain to attempt it.

I can only conclude that Mr. Constable was a man of few words!

Several key points are worth noting in Arden and Constable's experiments: the Leyden jar's central wire has several points of loose contact, including chain links and a sliding joint. The jar is large and is charged slowly, almost to the flash-over point. The open jar, the overhead window, and the fact that the visible fireball was formed early in the tests, but not later, suggests that a film of moisture in the glass jar and the central glass tube is the most important requirement. In the historical evolution of the Leyden jar, the early form included no lid. Natural philosophers soon discovered that by blowing into the jar, the moisture would greatly increase the storable charge.

Many theories have been advanced to account for the large amounts of energy and long lifetimes of these electric entities (usually 1 to 5 seconds). Fireballs have been known to cause major damage to tile roofs and chimneys, bend heavy iron gates and door hinges, bring a barrel of water to its boiling point in a short time, and bore small holes through granite blocks. Some of these points are described in Brand's report of 1923. None of the theories mentioned in these two sources requires a major shift in fundamental physics concepts.

Theoretical implications

I personally feel these explanations do not account for the enormous energy content in a small space. The three artificial electric-entity productions featured here involve no large power requirements to initiate a formation.

Following are two unorthodox theories, seldom seen:

1. In line with Gustave Le Bon's ideas on the universal dissociation of matter, a fireball would be seen as a slow energy-conversion process, in which elemental matter (water molecules, for example) yields its intrinsic potential energy. When the quantity of moisture is slowly reduced to a low level, the fireball becomes starved out of existence. Le Bon pointed out the

large number of electrons bound in a single gram of water. Should these electrons be freed to appear as charge, an enormous 96,000 coulombs of electricity would be produced. One coulomb applied to each of two spheres 1 meter apart represents an electrostatic force of 2 billion pounds! Normally, only tiny fractions of a coulomb are found in nature.

2. The second view involves a return to our discussion of gravitation. In this theory, an energy conversion occurs that disturbs the "concealed or hidden" motions in space (hypothesized by Heinrich Hertz in 1899 to account for the storage of potential energy).

Also, odd gravity anomalies are associated with electric entities, specifically, heavy, fragile objects that fall without being broken. This action implies that the basic properties or constants of space itself have been altered. Since the true nature of electrification is concealed at the molecular level, perhaps electrification produces disturbances in the incessant, hidden motions of space. These disturbances are manifested as heat, light, and mass motion—one form of which is the appearance of ball lightning.

These two views, of course, are quite unsettling because they require a major paradigm shift in fundamental physics concepts and an enlargement of our scientific foundational principles. Some physicists have admitted that the formation of ball lightning in metal enclosures, such as airplanes, raises the question of how such large energy densities in a small space are maintained (see Nature, vol. 224, 1969, p. 895, for example). The metal enclosure ensures that fireballs are not supplied externally with electromagnetic energy, as some theoretical physicists have imagined. We have, in ball lightning, a means by which energy is extracted from the nearby quiescent environment and manifests itself as heat, light, and mass motion. If this conclusion is justified, then it would be a natural case of negative entropy; that is, energy flowing "uphill," not downhill, as is required by the Second Law of Thermodynamics.

One article in favor of this idea is "The Second Law of Thermodynamics and the 'Death' of Energy, with Notes on the Thermodynamics of the Atmosphere," by Charles P. Steinmetz from the *General Electric Review* (July 1912). The abstract of the article follows:

Expressing the second law of thermodynamics in the words: "Without expenditure of some other form of energy heat flows only from higher to lower temperature," the author shows that the logical sequence from this is the conclusion that eventually all energy transformation will stop, i.e., all motion will cease and the universe will be dead. The conclusion is not a reasonable one and the author sets out to disprove the general applicability of the law. Adopting as his line of reasoning the thermodynamics of gases, he shows how, attending the escape of molecules from the attraction of earth into cosmic space, there is a heat energy flow from a temperature of 10°C to one of 60,000°C. Even within the earth's atmosphere, and without considering what happens in cosmic space, he shows that there is a transference of heat energy from lower to higher temperatures, or rather against the thermodynamic temperature equilibrium; and leads us to the conclusion that this law of thermodynamics is not of universal application, but applies only within the limited range of thermodynamic engines, from which it has been derived.

In addition to Dr. Steinmetz, physicists James Clerk Maxwell, Thomas Preston, and Lucien Poincare held a similar view of the Second Law of Thermodynamics.

A thought-provoking implication presented itself when I reflected on Arden and Constable's simple experiment, performed during the latter part of the eighteenth century. With their homemade frictional generator and Leyden jar, they succeeded in producing a phenomenon that today's government research centers have failed to duplicate using the best high-powered generators and large financial investments.

Extrapolating from this paradox, it now appears that it is possible to produce quite anomalous results by employing a larger number of principles through which nature operates. The experimental lab should duplicate natural environmental conditions; sterility and uniformity are often barriers to the discovery of new natural laws.

The application of this philosophical approach could result in a great simplification of our technology, making it more reliable with less waste and pollution as by-products. These possibilities require a greater resiliency and a willingness to think in different modes, some of which have not been emphasized in our present educational system. Fortunately, the spirit of inquiry and creativity is innate in each new generation of children.

Addendum

Recent experiences of interest to ball lightning experimenters

Since this chapter was first written, additional information has been gathered on unusual electric discharges. I include historical details of very early experiments long since forgotten and bring these up to date, provide new details on the Arden and Constable experiment and two eyewitness accounts, and discuss what experiments they suggest.

Expansive power of confined electric discharges

From 1800 to 1815, natural philosopher Mr. DeNelis of Holland experimented by exploding wires in metal cylinders filled with liquids. The results impressed two other experimenters, George Singer and Andrew Crosse, so they repeated these experiments and reported their results in the leading science journal *Philosophical Magazine* in 1815. Using two frictional electrostatic cylindrical glass generators, one 52 inches in circumference and the other 40 inches in circumference, they charged a battery of 50 Leyden jars totaling 75 square feet of coated surface. It required 2 minutes to reach a full high-voltage charge. Thick cylinders of different metals, including bismuth, zinc, tin, lead, iron, copper, and brass, were used. Each cylinder measured ½ inch in diameter × 2 inches tall with a bore of 0.20 inch × 1½ inches deep. A steel needle $\frac{1}{40}$ inch thick × 3 inches long was placed in the center of the bore. The end of the needle was prolonged by a short lead wire $\frac{1}{100}$ inch thick that touched the bottom of the bored cylinder. This needle was kept centered with two wax beads that electrically insulated the needle and wire from the walls of the cylinder. Besides the needle, each cylinder was filled with either water or olive oil. The cylinder being tested was placed in a wood box to contain the explosion; the bottom of the cylinder was connected to the outside of the Leyden jar battery, and the needle

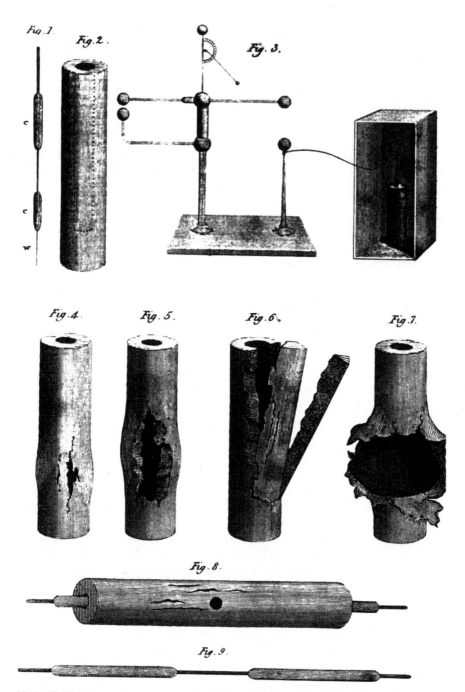

20-11 DeNelis' experiments on metal cylinders. *Philosophical Magazine,* 1815 (article by George Singer and Andrew Crosse).

completed the circuit to the inside coating. The resulting discharge exploded the lead wire that was communicated to the cylinder by compression of the confined liquid. The enormous force produced is evident in Fig. 20-11. Depending on the metal chosen, it required from 1 to 15 discharges to produce cylinder rupture. DeNelis also tested a thick 2-inch diameter cylinder of small bore outdoors. Upon explosion, the contained water was blown to a height of 40 feet!

Many years later Dr. Francis Nipher continued these impressive experiments and described them in his book, *Experimental Studies in Electricity and Magnestism* (1914). He concluded that when the thin lead wire at the bottom of the cylinder is suddenly drained of electrons (by joining it to the positive terminal of a capacitor bank, the atoms of lead repel each other, and the lead becomes explosive and acts hydraulically on the liquid as a compressional wave). Heating effects alone could not account for the enormous and sudden force. Dr. Nipher found the effects more marked with a negative discharge through the wire than with a positive discharge, the first giving a compression wave and the last giving a rarefaction wave. He compared the results of his capacitor discharge with that produced from a 250-volt dynamo and a 600-volt line from the city power plant. The lead wire was fused and converted to fine powder without any explosive effect.

Bringing these experiments up to date, Peter Graneau has developed a process called *water-arc explosions* that blasts a small slug of water out of a modified gun barrel using a high-energy pulse-discharge capacitor. Average water velocity was calculated to be 307 meters per second in one test. Dr. Graneau, like Dr. Nipher, also concluded that superheated steam cannot be the cause of the large force produced; the water never reaches the boiling point, and the explosions are cold. Some of his results were published in *Physics Letters A* (28 July 1986).

Explosive power of water films

While experimenting with meandering sparks, I decided to make a series of hurdles of baffles by cutting parallel grooves into an acrylic plate and placing a microscope slide in each groove. A second slide was placed with a film of water in between next to each slide, creating a sandwich. Figure 20-12 shows the setup. I wanted to see if a Leyden jar spark would pass from terminal to terminal over this barrier or meander through the series of water films. A large Leyden jar having a capacity of 0.0008 microfarads was charged to approximately 200,000 volts with the large Wimshurst generator and discharged across the arrangement of slides from ball to ball. I felt the blast in my face and was quite surprised to see that most of the slides were blown to tiny bits that spread out to a distance of about 15 feet radius. The glass dust was so fine that an eye loupe was needed to see that it was glass. On examining the acrylic plate, I found the only glass remaining intact to be that left in the plate grooves! I immediately remembered the earlier work of Dr. Nipher. These explosions involving water are more pronounced when the capicitance and inductance are small to keep the time constant short; this increases the suddeness of the discharge, and the force is magnified.

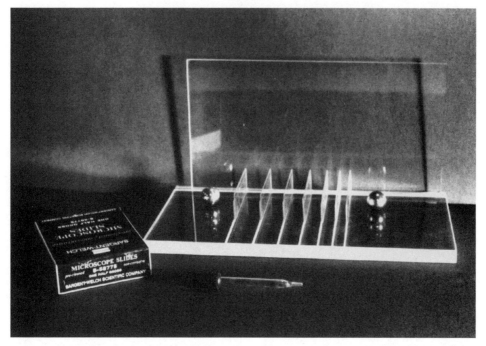

20-12 Setup for exploding glass slides using discharge through water films.

How are these experiments of importance to the subject of ball lightning? They indicate that large forces can be developed when lightning discharges pass over moist porous surfaces such as sandstone, wrought iron, or chimney brick. Jets of water combined with metals or hydrocarbons could form toroidal vortices similar to Hill's vortex in line with that proposed by Karl Wolf in 1915. However, the rotational speeds within these vortices may create new secondary effects not normally found in hydrodynamics as well as increase the lifetime of the toroidal entity.

Revisiting the Arden and Constable Experiment of 1757

This earlier mentioned experiment (Fig. 20-10) is appealing because of its long-lived spinning sphere, the elegant simplicity of the circuit, and the very small power required. Dr. David J. Turner, writing in *Physics Reports* (1998), points out some of the many variables to be encountered in reproducing the Arden and Constable experiment. In July 1999, I experimented with a similar circuit to that original drawing published by Cavallo in 1795. The sketch shows that the two philosophers were testing a lightning rod on the model house FF using a Leyden jar of about 3- to 4-gallon size to get a hotter spark. The arrangement would act as a damped series resonant circuit having small inductance and a capacitance of about 0.001 to 0.003 microfarads depending on the height of the coating and dielectric constant of the glass. Note that while brass chains complete the circuit between the jar and the house, the jar's outside coating is imperfectly grounded through the wooden table to the floor. This means that the jar's outside coating could "float" above ground potential. Since only glass cylinder and glass disc frictional generators were produced back in the eighteenth century, I concluded that the charge delivered to the inner terminal of the Leyden jar must be positive, not negative. If Mr. Arden and Mr. Constable experimented in winter with a window close to their frictional generator, then condensed moisture in the glass jar and on the control glass tube and brass chain was another factor to consider. I concentrated my experiments on those factors I could duplicate: varying degrees of moisture for the dielectrics and the sliding brass rod and chains shown. I used a 24-inch sectorless Wimshurst and 5-gallon bucket capacitors. Inner linings for the capacitors were either sheet lead, copper, or pure powdered iron. I fashioned short chains 4 inches long with links of 16-gauge iron, brass, or copper wire, the links being either open or soldered with a butt joint. I used these both as shiny metal for low resistance (less than 1 ohm) and also treated chains with acid and flame to give oxide coatings; oxide coatings could raise the resistance of the 4-inch chains to 15 megaohms. These chain lengths were joined to the bottom of the central brass rod and touched the inside bottom of the bucket coating. Either glass or plastic tubes were slipped over the rods as shown in the original drawing. Since Mr. Arden states that the fireball was first seen coming up the central rod conductor and glass tube, I concentrated on this part of the circuit. Soot-coated chain also was tried, and I used tap water to moisten the inside and outside surfaces of the bucket capacitor as well as the tube, rod, and chain. In addition, ice was placed in the bottom of the bucket, or alternatively, the glass tube and chain were cooled with dry ice so that frost was formed on them when they were reinstalled. None of these variations produced any unusual discharges.

I finally concluded from these new experiments that the fireball likely was formed as a result of the manufacturing defects peculiar to the crown glass jar-making process. (I assume that such jars are no longer made.) I did not do these experiments with Pyrex glass containers because this dielectric does not absorb water or permit spreading water films. Older glass would have variations in thickness and contain swirls, bubbles, pores, and furrows.

Those trying to reproduce this eighteenth-century experiment should know that the Burton Constable museum near Hull in Yorkshire, England, still has in its collection scientific instruments owned by William Constable.

Among the instruments listed is a glass wheel frictional generator believed to have been built by the well-known British instrument maker Benjamin Cole and delivered to Mr. Constable in early February 1757, about 10 months before the legendary experiment took place. Two large Leyden jars (in their original case), "silvered" inside and out, were preserved. These were delivered by Benjamin Cole in early July 1757. The word *silvered* for the coating is most likely a misnomer because in studying the history of mirror making, or "looking glass" making, as it was called in the eighteenth century, I found that true silvering with that metal onto glass began about 1840. Before this, the process was called *amalgamation*, that is, an alloy of metals was used with a large percentage of mercury, which is liquid at low temperatures. One such formula for silvering glass globes includes $\frac{1}{3}$ ounce each of lead and tin, $\frac{1}{2}$ ounce of bismuth, and 5 ounces mercury. The new coating was silvery-white in appearance, whereas true "silvered" glass had a slightly yellow cast in appearance. (Those using amalgams in their looking-glass businesses would have short careers because of the very toxic vapors from the mercury and lead!) However, it does show that William Constable spared no expense on his experiments, since most Leyden jars of his day were economically coated with lead or tin foil.

Recommendations for future experiments

Researchers attempting to duplicate the Arden and Constable experiment should concentrate on the microstructure and chemistry of glass surfaces coated with the same heavy-metal alloys used in the 1750s in England. Among the factors worthy of investigation I would include the crown glass process; jars made this way would have a characteristic "pontil mark," also called a *bull's eye,* in the center of the jar's bottom with concentric furrows or ridges expanding outward and up the sides. Various methods for cleaning the glass for coating should be tried, including acid baths, since making the metallic coating stick when charged was important to instrument makers. Since the jars had large openings, amalgam alloy coatings could be rubbed down and polished by hand. Mr. Arden's description mentions first seeing the red fireball come up from the center of the jar and that the brass chain touching the bottom of the jar was of short length. Because of the low melting point for the metallic coating, metallic vapors plus condensed moisture might be essential in the formation of electric entities. Pores and capillaries in the glass surface could become sites for DeNelis jets, which would help a tornoidal vortex to form in the base of the Leyden jar. In closing, the preceding descriptions are lengthy, but electrical experimenters wanting to replicate this most unusual work so early in the history of electrical science need to give careful attention to historical research of the methods used by instrument makers in the eighteenth century.

New experiments suggested by three eyewitness accounts

Of great importance to electrical experimenters are those eyewitness accounts of electric anomalies which suggest how to set up circuits in novel ways.

The first report was given to me by John Monteith, who at the time of the event (1970s) was living 40 miles north of Seattle, working as a telephone installer. Mr. Monteith was standing in his front room after dark when a thunderstorm developed. He saw a flash but heard no thunder. In the same instant he saw a bright yellow "disc" roll out of one electrical outlet, roll across the nylon carpet, and enter another outlet on the opposite side of the room, a distance of 12 feet. He assumed it was a disc not a ball because it reminded him of a circular saw blade. Size was about 3 to 5 inches in diameter, and the transit time was about 3 seconds. No smoke or sound was produced by this entity, but the smell of ozone was noticeable. No scorch marks were left on the nylon carpet, no soot marks or damage was done to either outlet, and the circuit breakers were not tripped.

The second event happened to me while I was experimenting with the large sectorless Wimshurst with 24-inch discs. The drive motor was plugged into an overhead duplex receptacle shared by a two-wire cord running to a wall-mounted Honeywell switching relay (model #RA89A). This relay was located about 10 feet away from my Wimshurst. Over the course of several months, I counted seven unusual discharges coming from near the relay. Most of these were seen out of the corner of my eye because I was focused on the Wimshurst with my back to the relay.

However, I had to admit a cause-and-effect relationship when on two discharge occasions the Wimshurst drive motor slowed to a stop and it was found that the motor was burned out! These ac gearhead motors were identical and taken from discarded copy machines. On dismantling each motor, I found an open circuit in the field windings. I then discovered that wall discharges occurred when small ¼-inch sparks passed from the bead chain connecting the outer Leyden jar coatings and the ungrounded motor casing; if allowed to continue, the unusual discharge increased in intensity, and the second or third such would burn out the motor. Two of the last discharges stand out because by now I was focused on the relay, not the generator. From the ac cord leading up to the base of the switching relay, a shower of bright orange-yellow sparks shot out and downward to the concrete floor about 3 to 4 feet from the wall. The last discharge was oval-shaped without sparks, had a pink center surrounded by red, and was about ¾ inch across. On inspecting the cord going into the relay, I could see that most of the insulation at that point was gone from the solid copper wires, and the wires were between ¹⁄₁₆ and ⅛ inch apart and blackened with soot. However, in none of these seven discharges was the 20-amp circuit fuse blown or any damage done to the relay internally! I did manage to test all wires running from the generator motor to the overhead outlet and from there along the cord to the Honeywell relay; I passed an uncharged electroscope along this path with the Wimshurst running and found no charge built up.

The third event was the "laboratory illustration of ball lightning" by A.T. Jones in 1910 described earlier in this chapter, in which a copper wire was short-circuited with 110 Vac and produced a small fireball. Since there is no record of this result being produced again in the many years since, the two previous accounts offered the possibility of an electrostatic spark causing Jones' short circuit by connecting the ac-energized copper wire to ground.

Hybrid circuits combining electrostatic sparks with ac conductors

The three previous accounts encouraged me to set up circuits in which a spark from a Leyden jar could be passed onto an energized 110-Vac conductor in close proximity to a parallel ground wire; the resulting discharge would be compared with simply touching the two wires together to create a short circuit; the ac line was protected by a 20-amp circuit breaker.

Figure 20-13 shows one setup. Two annealed black steel wires are run parallel to each other about 1 inch apart; each wire is 16 gauge by 3 feet long, and the pair is supported at each end with acrylic rods screwed to a plywood base plate. At one end of the wires the top wire is joined to the energized side of an ac plug, and the bottom wire is connected to ground. Near the same end of the conductors, a ¾-inch ball terminal is placed 1¼ inches above the top wire, the terminal connecting with the inside of a 5-gallon bucket capacitor. A capacitance meter rated this at 0.0008 microfarads. The capacitor is charged slowly with the small Wimshurst generator, but a Van de Graaff could be substituted. (Electrostatic generators are preferred because their drive motors are isolated from the high-voltage output as well as from voltage spikes sent back through the ac conductors.) Near the other end of the wires, I pinch the wires together to give a short gap of 1 to 2 millimeters. Most of the anomalous discharges occur at this gap. Since the capacitor discharges randomly, the big "events" appear to occur when the ac output is at a specific point in its cycle. The unusual images, captured on videotape, include vertical blue-white jets, comet-shaped, with long narrow tails, and spheres with sizes ranging from about 6 inches in diameter upward with colors including white, blue, amber, or yellow. When the wires at the short gap are contaminated with hydrocarbons such as charcoal dust and water, larger events take place. Clusters of white balls form, each approximately ⅜ to ½ inch in diameter with lifetimes ranging from 30 milliseconds to 0.8 second. Each ball falls under the influence of gravity and is extinguished as its diameter shrinks.

Larger events are also more pronounced if the capacitance is kept small. When I doubled the capacitance, the discharges were smaller and less frequent. At present, I have concluded that having a small time constant gives a more abrupt shock to the circuit, which is favorable. When the voltage was stepped up to 800 Vac at 100 milliamps, these effects were greatly reduced; I concluded that these events depend on the instantaneous *current* in the conductors rather than the instantaneous *voltage*. What is most amazing is that none of these discharges tripped the 20-amp in-line circuit breaker because they are such sudden short circuits! Also, there is very little evidence of melted wire at the short gap. As a comparison, when the top ac-energized conductor and the bottom ground wire are pinched together to make contact, a small spark is produced, the wires fuse together, and the circuit breaker trips.

Figure 20-14 shows a single video frame in which a blue-white ball approximately 6 inches diameter was formed with an indistinct outer edge. The printed image revealed a closely grouped cluster of small white spots as a nucleus. Even though the unusual discharges from this hybrid circuit cannot be called ball lightning because of their short duration and their being influenced by gravity, they still provide useful information. A more complicated internal structure is suggested by the very luminous spotted nucleus. Although power lines are absent in many eyewitness accounts, we still cannot rule out strong earth currents inducing voltages in grounded metal conductors and moist semiconductors in contact with the ground, which then interact with an electrostatic spark to form a hybrid circuit.

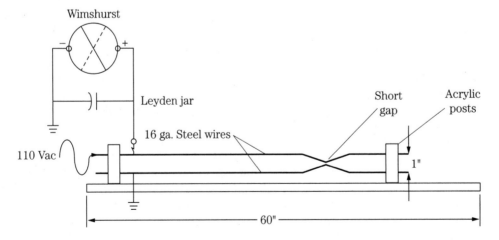

20-13 Electrostatic ac hybrid circuit.

20-14 Single video frame of spherical discharge, 1/30 second. Note bright spots in nucleus.

Fortunately, the experiment just described is simple to set up and document with video, and the power requirements are very modest. I find it helps to balance reason and logic with intuition and a careful observation of nature's examples.

21
CHAPTER

Some philosophical conclusions and insights

As interesting as foregoing experimental avenues might seem, the most important part of the book is discussed in this chapter. That is, how we see, think, and reason on scientific subjects is more important than project plans. The character of our science and technology, especially as it impinges on the environment in the looming specter of toxic wastes and the ravaging of earth's resources, should impel us to look for lasting solutions.

Seldom do we go far enough to reexamine how we think through solutions without creating new problems in their wake. I previously mentioned the importance of intuition and visualization as elements of the innovative process. The three main elements that are important in creative thinking are: visualization, intuition, and qualitative anatomy.

Visualization

By *visualization,* I am not referring to a self-hypnotic state, but rather that ability to form models in thought, as for example, the Le Sage concept of gravitation.

In Newton's time, there was an increasing fascination with abstract mathematics as a method of "seeing" scientific principles. By the turn of the twentieth century, visualization as a tool had fallen into disuse.

In one sense, abstract mathematical reasoning is identical to taking a subway in a large city; we have no way of knowing whether a far shorter route exists. Now, please do not conclude from what I have said that mathematics and its offspring, statistical and computer analysis, have no place in scientific innovation. Rather, they must be balanced with an equal emphasis on the other ways of "seeing" the way in problem-solving.

Generally, mathematics should be used to find exact solutions only after we have diligently developed working mental models to explain phenomena. Finding these mental models requires participating in concrete laboratory experiments. The use of math without physical experiments leads to the construction of "air castles," which have no firm foundation in reality. Ultimately, you must visualize the principle under consideration before you can use the full powers of your intellect to find an innovative solution.

Intuition

Intuition is the ability to reach a solution without recourse to reasoning or inference. It is an inner knowing or deep feeling. Even though intuition is innate in children, it usually is suppressed by the educational system.

Although logical, linear thinking proceeds from point A to point B to point C, intuitive thinking leaps from A directly to C. Sometimes intuition is thought to be the exclusive domain of women's thinking; however, this idea is completely unfounded. For example, inventor Nikola Tesla, who pioneered much of our existing electrical distribution system, often had flashes of insight that supplemented his mathematical reasoning.

The long history of inventiveness in the United States has not taken gradual steps, but usually has proceeded by intuitive leaps. For example, powered flight and the development of the telephone come to mind.

Now consider some aspects of contemporary technology and compare them to what our intuitive sense indicates they should be. Technical devices such as motor vehicles could be thought of as noisy, considerably inefficient, polluting, and complex. What should they be? In this case, we can arrive at the answer almost by inversion: silent, efficient, nonpolluting, and very simple (reliable). So now we can assess what are truly innovations in transportation. Normally, intuition answers our questions of what should be.

Qualitative anatomy

Qualitative anatomy is a method of seeing that involves a resolution of the object under consideration into one-word qualities that characterize the object. The following are three everyday examples of qualitative anatomy. The words in parentheses simply explain our quality choice.

Pencil
Communication
Rigidity

Utility pole
Communication (supports phone lines)
Rigidity
Uniformity
Sterility (nonliving)
Safety (supports and protects power lines)
Supply (links to electrical energy)

You would hardly expect to find a landscape painter getting ecstatic over the prospect of painting a panorama of utility poles. Why? Because uniformity and sterility are not attractive qualities in the landscape.

Tree
Communication (for flocks of birds)
Rigidity (main trunk and limbs)
Permanence (settled, rooted)
Protection (for birds and squirrels)
Individuality
Resilience (twigs and leaves)
Supply (leaf nutrients and lumber)
Comfort (natural air-conditioner)
Beauty
Alive

Note that trees are a "higher" expression because they include more qualities; the lesser (pencils and poles) can only come from the greater (trees).

Qualitative anatomy is not a trivial pursuit. It exercises those mental faculties that we need in to order to "see," just as poets and painters see their surroundings. With continued practice, this tool becomes very helpful for assessing the value of both objects and abstract concepts, such as beauty or work. Qualitative anatomy also assists our experimental work.

Now, combine these three aspects of creative thinking into a mental juggling act. The list of qualities you arrive at will be strongly influenced by your reference point. For example, your point of view might be influenced by possible social or environmental impacts. The reference point concept can be illustrated by contrasting the nature of progress, as it is now, with what your intuitive sense believes it should be. The viewpoint is that of social impact; how progress relates to people.

Progress

Contemporary	*Intuitive sense*
Complexity	Simplicity
Specialization	Diversity
Self-pollution	Nondestructive
Material acquisition	Spiritual wealth
Rigidity	Resilience
Without identity	Individualism
Narrowing of skills	Multiple talents, skills
Sterile uniformity	Spontaneous variety
Vulnerability	Stability, strength
Centralized control	Local autonomy
Transience, homelessness	Settled, rooted
Convenience, ease	Nurturing, individual responsibility
Insecurity, stress	Peace of mind
Frenzy, rush	Poise, dignity

When we hold these contrasting views of progress in mind, we can then assess which new developments in technology truly lead us into a better world. Science and technology must be responsible to social needs, and to how their products affect the character of people.

If we would give serious attention to these modes of thinking, it would enable the rising generation of students to work with qualities as precisely as they deal with mathematics. Qualification would then reach an equal standing with quantification.

Some useful surroundings

A natural setting is helpful for reflective thought. Surrounded by nature's examples, we are more likely to recognize different principles that operate in the environment. Silence and a relaxed, tranquil attitude are essential.

At first, strongly focus on the technical problem to be solved. If answers and insights do not come quickly, put the task out of your thoughts entirely. The worst mistake is to concentrate on a problem until it is solved. This is a weakness of the "think tank" or "brain trust" approach; it usually leads to a myopic, hypnotic state, in which you cannot recognize new principles that will greatly simplify the solution.

One extreme case of myopic thinking occurs to me. In 1962, I was working on a mechanics problem. After two days of thinking and drawing, I still had no answer. Eventually, I forgot about the subject until I came across it again in 1986. I briefly reviewed some aspects of the mechanics involved and made a list. Then one night, just before falling asleep, a mental picture flashed through my mind, showing the mechanism in action. I jumped up and drew a sketch, and it did satisfy all the requirements as a solution! Apparently, an incubation period is sometimes necessary. Always have a pencil and notepad on your nightstand so you don't lose these flashes of insight.

To take part in a new renaissance, men and women need to avoid becoming specialists. We should adopt the multidisciplined approach with some knowledge in several different areas. The development of skills in the industrial arts is also essential in experimental research.

Orthodox science too often takes on an air of omniscience; that is, that all the basic knowledge we have is all there is. This position is further complicated by its companion attitude—infallibility. Orthodox scientists often believe that there is only one way and we have never taken wrong or dead-end paths.

Especially since the development of the atomic bomb in the 1940s, science and technology have taken on what might be called a "swaggering arrogance"; this attitude has no place in the science of the twenty-first century. Nature is not to be dominated. Its many principles should be copied in the applications of our sciences and technology.

Any scientific body of knowledge that has excluded natural anomalies because they do not fit with accepted notions, must not only be suspect, but be incomplete. Most of the natural anomalies found in scientific journals over the past two centuries still have not been classified and catalogued. The task for those researching the full history of scientific discovery remains awesome.

I have tried in this book to select anomalies and innovations found before the turn of the twentieth century, in the hope of bringing respect and admiration to the spirit of inquiry in those earlier times. We have much to relearn and reincorporate into our own time. We must blend it with the best of what we have to offer.

In view of this, it appears that the real "high frontier" is still very much on earth, rather than in space. The challenge is to not only reexamine the way we think, but to review the hidden assumptions in the foundation on which today's science rests.

Happy experimenting!

Appendix

Cataloguers of nature's anomalies

The Source Book Project, P.O. Box 107, Glen Arm, MD 21057.
   Provides high-quality books on anomalies and publishes a newsletter *Science Frontiers*.
Pursuit (Journal), c/o Edward Brother's, Inc., 2500 S. State St., Ann Arbor, MI 48104.
   A journal dealing with unexplained natural phenomena.

Suppliers

AIN Plastics, Inc., 300 Countyline Rd., Bensonville, IL 60106.

The Al-Chymist, 17130 Mesa St., Hesperia, CA 92345.
   Supplies chemical and lab equipment.

American Science & Surplus, 3605 Howard St., Skokie, IL 60076.
   Excellent supplier of surplus equipment, including 12-volt motors, ball bearings, optics, flash rocks, low-voltage rectifiers, transformers, and science fair items.

Antique Electronic Supply Co., 6221 S. Maple Ave., Tempe, AZ 85283.
   Supplies high voltage tube diodes.

C and H Sales Co., 2176 E. Colorado Blvd., Pasadena, CA 91107.
   Sells motors, rheostats, transformers.

Cherry Tree Toys, Inc., P.O. Box 369, Belmont, OH 43718.
   Supplies wood balls for Wimshurst generator construction.

Coltene-Whaledent, Inc., 750 Corporate Dr., Mahwah, NJ 07430-2009.
   Supplies dental dam.

Edmund Scientific Co., 101 E. Gloucester Pike, Barrington, NJ 08007-1380.
  Sells lenses and general science fair materials.

Elemental Scientific LLC, P.O. Box 571, Appleton, WI 54912-0571.
  Supplies sulfur, paraffin oil, lamp black, and lab equipment.

Herbach & Rademan Co., 353 Crider Ave., Moorestown, NJ 08057.
  Sells shaft extenders, ball bearings, motors.

Arthur Harris & Co., 210 N. Aberdeen St., Chicago, IL 60607.
  Supplies #304 stainless-steel float balls for high-voltage terminals. Has minimum
order requirement.

Howee's, Inc., 2220 S. Prosperity Ave., Joplin, MO 64801
  Supplies wooden balls for Wimshurst handle socket.

Klaus Radio, Inc., 8400 N. Allen Rd., Peoria, IL 61615.
  Supplies high-voltage varnish, knife switches, and rectifiers.

McMaster-Carr Supply Co., P.O. Box 4355, Chicago, IL 60680-4355.

N.T.E. Electronics, Inc., 44 Farrand St., Bloomfield, NJ 07003.
  Makes high-voltage diodes used in industrial and microwave ovens. Write for
your local distributor.

Science First/Morris & Lee Co., 95 Botsford Place, Dept. A, Buffalo, NY 14216-2696.
  Has supplied affordable Van de Graaff generators and accessories to the public
since 1960.

Small Parts Inc., P.O. Box 4650, Miami Lakes, FL 33014-0650.
  Good hardware supplies for inventors.

Surplus Center, P.O. Box 82209, Lincoln, NE 68501-2209.
  Supplies 12-volt electric motors, transformers, rectifiers.

Swift & Sons, Inc., 10 Love Lane, P.O. Box 150, Hartford, CT 06141-0150.
  Supplies gold-leaf books for electroscopes.

Swordmark Co., P.O. Box 49592, Atlanta, GA 30359.
  Supplies sealing wax sticks for electrophorus cakes.

Thomas Register. Consult this publication at your local library for the following:
Steel and brass shim stock (for neutralizers and Leyden jars), wax, carnauba and
gum, ester (for electrophorus cakes).

United States Plastics Corp., 1390 Neubrecht Rd., Lima, OH 45801.
  Supplies general acrylic plastics and clear Butyrate shipping tubes.

Useful books

De Cristoforo, R.J. *De Cristoforo's Complete Book of Power Tools*. Popular Science
Publishing Co., NY, 1972.
  Covers woodworking techniques using simple, inexpensive jigs and fixtures.

Stong, C.L. *The Scientific American Book of Projects for the Amateur Scientist.* Simon & Schuster, NY, 1960.

Strong, John. *Procedures in Experimental Physics.* Prentice Hall, NY, 1938.

Taylor, Charles. *The Art and Science of Lecture Demonstration.* Adam Hilger, PA, 1988.

White, Harvey E. *Modern College Physics*, 4th ed. D. Van Nostrand Co., NJ, 1962.

Research bibliography (by subject)

Electronic equipment protection

Greason, William D. *Electrostatic Damage in Electronics: Devices and Systems.* Hertfordshire, England: Research Studies Press, 1987.

Sclater, Neil. *Electrostatic Discharge Protection for Electronics.* Blue Ridge Summitt, PA: McGraw-Hill, 1990.

Frictional electric generators

Dibner, Bern. *Early Electrical Machines.* Norwalk, CT: The Burndy Library, 1957.

Frick, Joseph. *Physical Technics.* Philadelphia: Lippincott & Co., 1878.

Hackmann, W.D. *Electricity From Glass: The History of the Frictional Electric Machine, 1600–1850.* Netherlands: Sitjhoff and Noordhoof, 1978.

Harris, William Snow. *A Treatise on Frictional Electricity.* London: Virtue and Co., 1867.

Weinhold, Adolf. *Introduction to Experimental Physics.* London: Longmans, Green and Co., 1875.

"Winter's Electrical Machine." *English Mechanic*, vol. 2, Oct. 20, 1865.

The influence machine (and modifications)

British Patent No. 22, 731, Tudsbury.

Chaumat, Henri. "The Chaumat Electrostatic Machine." *Societe Frances Des Electriciens*, bulletin #1, p. 673+, July 1931. (In French.)

Felici, Noel J. "Electrostatic Generators." *Direct Current*, vol. 1, June 1953.

Gray, John. *Electrical Influence Machines: Their Historical Development and Modern Forms.* New York: Whittaker & Co., 1903.

"Influence Machines." *The Electrician*, vol. 35, London, 1895.

Johnson, Valentine E. *Modern High Speed Influence Machines*. London: E. and F.N. Spon, 1921.

Jolivet, Pierre. *Sur une novelle machine Electrostatique a Influence*. Paris: R.G.E., 1953.

Marinov, Stefan. *The Thorny Way of Truth. Part 5*. Explores the history of European electrostatics up to recent developments, 1989.

"The Wommelsdorf Condenser Machine." *Journal of the Rontgen Society*. Jan. 1914.

Thompson, Silvanus. "The Influence Machine From 1788 to 1888." *Journal of the Telegraph Engineers*, vol. 17, Silvanus P. Thompson, London, 1888.

U.S. Patent No. 882,508 and 1,071,196, Wommelsdorf.

U.S. Patent No. 634,467 and 720,711, Lemstrom.

U.S. Patent No. 937,691, Baker.

U.S. Patent No. 821, 902, Todd.

U.S. Patent No. 1,109,205, Dempster.

General electrostatics

Craggs, J.D. and J.M. Meek. *High Voltage Laboratory Technique*. London: Buttersworths Scientific Publ., 1954.

Heilbron, J.L. *Electricity in the 17th and 18th Centuries: A Study of Early Modern Physics*. Berkeley, CA: University of California Press, 1979.

Moore, A.D. *Electrostatics*. New York: Anchor Doubleday & Co., 1968. Reprinted in 1997 by Laplacian Press, Morgan Hill, CA.

Moore, A.D. (ed.) *Electrostatics and Its Applications*. New York: John Wiley & Sons, 1973.

United States Patent Class Listing. "Electrical Generation—Induction Type," Class #310, Subclass #309.

This is a list of approximately 300 U.S. patents on electrostatic induction devices, including all influence machines. Available from the U.S. Patent Office.

Van de Graaff generators

Blewett, John and Livingston, M. Stanley. *Particle Accelerators*. New York: McGraw-Hill, 1962.

Gay, Loren W. "Lightning in Your Hand." *Popular Science*. October, 1946. (pp. 188–192)

Moore, Arthur D. *Electrostatics and Its Applications*. New York: John Wiley & Sons, 1973. (See Chapter 8, "Electrostatic Generators.")

Wierts, Derk. *A 5 MV Electrostatic Generator for Particle Acceleration*. Groningen, The Netherlands 1964.

The electroscope

Dolbear, A.E. *The Art of Projecting*. Boston: Lee & Shepard Publishers, 1877.

Le Bon, Gustave. *The Evolution of Forces*. London: Kegan Paul, Trench, Trubner & Co., 1908.

Le Bon, Gustave. *The Evolution of Matter*. London: Walter Scott Publishing Co., 1910.

Makower, W., and H. Geiger. *Practical Measurements in Radio-Activity.* London: Longmans, Green & Co., 1912.

Rosenblum, C. Frederick. "The Electroscope (Part 1)." *Journal of Orgonomy*, vol. 3, no. 2, Princeton, NJ: Orgonomic Publications, Inc., 1969.

Rosenblum, C. Frederick. "The Electroscope (Part II)." *Journal of Orgonomy*, vol. 4, no. 1, Princeton, NJ: Orgonomic Publications, Inc., 1970.

Royal Society of London, Proceedings. "C.T.R. Wilson Electroscope," vol. 68, p. 152, 1901.

Rutherford, Ernest. *Radio-activity.* England: Cambridge University Press, 1904.

Strong, John. *Procedures in Experimental Physics.* Prentice Hall, NY, 1938.

Tributsch, Helmut. *When Snakes Awake: Animals and Earthquake Prediction.* Cambridge, MA: MIT Press, 1982. Discusses electrometer anomalies and earthquake lights.

Weinhold, Adolf. *Introduction to Experimental Physics.* London: Longmans, Green and Co., 1875.

The Leyden jar

Heilbron, J.L. *Electricity in the 17th and 18th Centuries: A Study of Early Modern Physics.* Berkeley, CA: University of California Press, 1979.

Lodge, Oliver. *Lightning Conductors and Lightning Guards.* Macmillan & Co., NY, 1892.

"Liquid Condensers." *Popular Electricity Magazine.* Chicago, IL, 1912–13.

The electrophorus

Harris, William Snow. *A Treatise on Frictional Electricity.* London: Virtue and Co., 1867.

"Mushroom Generator." *Popular Mechanics.* New York, 1935.

Phillips, John. "On a Modification of the Electrophorus." *Philosophical Magazine*, series 3, vol. II, London, 1833.

Weber, Joseph. *Abhandlung Von Dem Luftelektrophor.* Ulm, Switzerland, 1779.

Research avenues

Electrostatic motors

Ford, R.A. "The Jumping Electrostatic Top." *Electric Spacecraft Journal*, issue 5, Jan–Mar. 1992, pp. 26–30, Leicester, NC.

Jefimenko, Oleg. "Operation of Electric Motors from the Atmospheric Electric Field." *American Journal of Physics*, vol. 39, July 1971.

Jefimenko, Oleg D. *Electrostatic Motors: Their History, Types, and Principles of Operation,* 1973. Electret Scientific Co., P.O. Box 4132, Star City, WV 26505.

Marinov, Stefan. *The Thorny Way of Truth, Part V.* Graz, Austria: East-West Publishers, 1989, pp 25–26.

Mascart, M.E. *Traite D' Electricite Statique*, vol. 1, Paris, France, 1876, Fig. 69, p. 179.

Pages, Marcel J.J., "Gravitation Antiponderale," *Revue Francaise D' Astronautique #3*, 1967, Fig. 8, p. 9.

Popular Electricity Magazine. "Static Electric Top." Chicago, IL, 1912.

Rho Sigma. *Ether-Technology: A Rational Approach to Gravity Control.* Lakemont, Georgia: CSA Printing & Bindery, 1977. New printing: Adventures Unlimited Press, Kempton, IL, 1996.

Electrohorticulture

Lemstrom, Selim. *Electricity in Agriculture and Horticulture.* London: The Electrician Printing & Publishing Co., 1904.

U.S. Patent No. 1,268,949, Fessenden.

U.S. Patent No. 1,952,588, Golden.

Electrotherapeutics and high-voltage humans

Becker, Robert and Gary Selden. *The Body Electric.* New York: William Morrow, 1985. Deals with general subject of bioelectricity.

Brown, E. Richard. *Rockefeller Medicine Men: Medicine and Capitalism in America.* University of California Press, 1979.

Coulter, Harris L. *Divided Legacy: A History of the Schism in Medical Thought.* vol. III. *Science and Ethics in American Medicine 1800–1914.* Washington, DC: McGrath Publishing Co., 1973.

"Electrified Convicts." *Electrical Experimenter*, vol. 8, p. 185, New York, 1920.

"Electrified Convicts." *New York Times*, p. 1, Monday, April 5, 1920.

"Extraordinary Case of Electrical Excitement." *Annals of Electricity, Magnetism & Electricity*, vol. II, London, pp. 351–354, 1838.

Strong, Fredrick Finch, M.D. *Essentials of Modern Electro-Therapeutics.* New York: Rebman Co., 1908.

Cold light

Birkeland, Kristian. *On the Cause of Magnetic Storms.* New York: Longmans, Green & Co., 1908.

Butman, Chester. "The Electron Theory of Phosphorescence." *Physical Review*, series 2, vol. 1, 1913.

Corliss, William R. *Lightning, Auroras, Nocturnal Lights and Related Luminous Phenomena.* Glen Arm, MD: The Source Book Project, 1982.

Crookes, William. *Radiant Matter.* Philadelphia: James Queen & Co., 1881.

Hansen, Steve. *An Experimenter's Introduction to Vacuum Technology*, 1995. (Available from the Bell Jar, 35 Windsor Dr., Amherst, NH 03031.)

Le Bon, Gustave. *The Evolution of Forces.* London: Kegan Paul, Trench, Trubner & Co., 1908.

"Lemstrom's Aurora." *Meteorological Magazine*, pp. 33–36, 51–55, London, April 1883.

"Lemstrom's Aurora." *Science*, n.s., vol. 4, pp. 465–466, 1884.

"Puluj: Phosphorescent Lamp." *Scientific American*, vol. 77, New York, 1897.

Fluorescence: Gems and Minerals under Ultraviolet Light. Manuel Robbins. Phoenix, AZ: Geoscience Press, Inc., 1994.

"Triboluminescence." *Journal of the Optical Society of America*, vol. 29, Wisconsin, p. 407+, 1939.

Electroaerodynamics

Dudley, Horace C. "The Electric Field Rocket." *Analog Science Fact & Fiction.* November 1960.

"Sonic Boom Experiments." *Product Engineering*, vol. 39, pp. 35–6, New York, March 11, 1968.

U.S. Patent No. 3,095,167, Dudley.

Valone, Thomas, ed. *Electrogravitics Systems*. Integrity Research Institute, Washington, DC, 1994.

Counter-gravitation

American Philosophical Society, Proceedings. Philadelphia, PA, years 1914–1929. See several articles on Charles F. Brush's experiments.

Brown, Thomas. "Controlling Gravitation." *Science and Invention.* New York, August, 1929.

"Can Electricity Destroy Gravitation?" *Electrical Experimenter*. New York, March 1918.

Hooper, William J. *New Horizons in Electric, Magnetic and Gravitational Field Theory*, Elsah, IL: Principia College, 1974.

"Nipher's Gravitation Experiments." *Transactions of the Academy of Science*, vol. 23, pp. 163–192+, St. Louis, 1916.

Niven, W.D., ed., *The Scientific Papers of James Clerk Maxwell*, vol. II, London: Constable & Co., 1965. See Maxwell's "Le Sage Theory of Gravitation."

"Piggott's Electro-gravitation Experiment." *Electrical Experimenter*, vol. 8, 1920.

Riemers, Chris. "Gravity by Corpuscular Radiation Pressure." *Speculations in Science and Technology* Vol. 18, 1995, pp. 205-215.

See, Thomas Jefferson. *WAVE THEORY! Discovery of the Cause of Gravitation!* Self-published, London, 1938.

U.S. Patent No. 1,006,786, Piggott; 3,518,462, Brown; 3,610,971; 3,656,013, Hooper.

Valone, Thomas. Ed. *Electrogravitics Systems*. Available from Integrity Research Institute, 1220 L St. NW, #100-232, Washington, DC 20005, 1994.

Unusual electrical discharges

"The Atomphysical Interpretation of Lichtenburg Figures and their Application to the Study of Gas Discharge Phenomena." *Journal of Applied Physics*, vol. 10, December 1939.

Barry, James D. *Ball Lightning and Bead Lightning: Extreme Forms of Atmospheric Electricity*. New York: Plenum Press, 1980.

Bertholon, Abbe. *De L'Electricite Des Meteors*, vol. 1, 1787.

Brand, Walther. *Der Kugelblitz* (Ball Lightning). Hamburg, Germany, 1923. (NASA Translation: NASA T.T. F-13 228, year 1971, Accession Number N71-18133).

Carpenter, Donald. *Ball of Lightning*. AIAA Student Journal, vol. I, no. 1. April 1963. pp. 25–27.

Cavallo, Tiberius. *A Complete Treatise on Electricity In Theory and Practice with Original Experiments*, vol. II, 1795.

Corliss, William R. *Lightning, Auroras, Nocturnal Lights and Related Luminous Phenomena*. Glen Arm, MD: The Source Book Project, 1982.

Corliss, William R. *Tornadoes, Dark Days, Anomalous Precipitation, and Related Weather Phenomena.* Glen Arm, MD: The Sourcebook Project, 1983.

De Villamil, R. *ABC of Hydrodynamics.* London: E. & F.N. Spon, 1912.

Egely, Gy. *Hungarian Ball of Lightning Observations.* KFKI-1987-10/D. Central Research Institute for Physics, Budapest.

Ehrenberg, W. "Maxwell's Demon." *Scientific American,* vol. 217, pp. 103–110, November 1967.

Fort, Charles. *The Complete Books of Charles Fort.* New York: Dover Publications, 1975.

Graneau, Peter and Neal. *Newtonian Electrodynamics.* River Edge, NJ: World Scientific Publ. Co., 1996.

"Laboratory Ball Lightning." *Science,* n.s., vol. 31, p. 144, 1910.

Leonov, R.A. *The Riddle of Ball Lightning.* T.T. #66-33253. August 1966 (U.S.S.R.).

Maxwell, James Clerk. *Theory of Heat.* New York: D. Appleton & Co., 1872.

Meaden, George Terence. *The Circles Effect and Its Mysteries.* Wiltshire, England: Artetech Publishing Co., 1989.

Persinger, Michael and Gyslaine Lafreniere. *Space-Time Transients and Unusual Events.* Chicago, IL: Nelson-Hall, 1977.

Poincare, Lucien. *The New Physics and Its Evolution.* London: Kegan Paul & Co., pp. 86–87, 1907.

Preston, Thomas. *The Theory of Heat.* New York: MacMillian & Co., 1904.

Ryan and Vonnegut. "Miniature Whirlwinds Produced in the Laboratory by High-Voltage Electrical Discharge." *Science,* n.s., vol. 168, pp. 1349–1351, June 12, 1970.

"Remarkable Effects of Lightning." *Scientific American Supplement,* vol. 57, p. 23679+, 1904.

Shoulders, Ken R. "Energy Conversion Using High Charge Density." U.S. patent 5,018,180, 1991.

Singer, Stanley. *Ball of Fire: Recent Research on Ball Lightning.* Springer-Verlag, 2001.

Space Energy Receivers—Power From the Wheelwork of Nature, 1993. Introduction to several inventor's attempts to harness atmospheric electricity. Send self-addressed long envelope to: Simplified Technologies, P.O. Box 2121A, Champaign, IL 61825.

Steinmetz, Charles P. *General Electric Review.* "The Second Law of Thermodynamics and the 'Death' of Energy, with Notes on the Thermodynamics of the Atmosphere." July 1912.

Turner, David J. "Ball Lightning and Other Meteorological Phenomena." *Physics Reports,* vol. 293, pp. 1–60, 1998.

"Unusual Damage By A Tornado." *Weather: A Monthly Magazine for All Interested in Meteorology.* Royal Meteorological Society, London, 1949.

Wainwright, Jacob T. "The True Second Law of Thermodynamics." *The Engineer.* vol. 113, pp. 658–9, London, June 21, 1912.

"Weisiger's Ball Lightning Experiments." *Electrical Experimenter.* February 1919.

Wolf, Karl. "The Nature of Ball Lightning." *Scientific American Supplement,* vol. 80, p. 54+, 1915

Exploding wire experiments

Braun, Ferdinand, "The Mechanism of Electrical Sputtering . . ." (in German), *Annalen der Physik*, 17, 1905, pp. 359–363.

Chace, William G. (ed.). *Exploding Wires*, vol. 1, New York: Plenum Press, Inc., 1959. Vols. 2 through 4 follow in later years. (Associate editor, Howard K. Moore.)

Chace, W.G., "A Bibliography of the Electrically Exploded Wire Phenomenon," *Geophysics Research Directorate*, Research Notes #2, ASTIA Doc. No. AD-152640, 1958.

Ford, R.A. "Exploding Wire Experiments." *Electric Spacecraft Journal*, issue 8, Oct–Dec. 1992, p. 28–32, Leicester, N.C. 46 Sunlight Drive, Leicester, NC 28748.

Graneau, Peter. *Ampere-Neumann Electromagnetics of Metals*. Palm Harbor, FL: Hadronic Press, 1985.

Nairne, Edward, "Electrical Experiments by Mr. Edward Nairne," *Philosophical Transactions.* Royal Society (London), vol. 64, pp. 79–89, 1774.

Nasilowski, Jan. "Unduloids and Striated Disintegration of Wires," pp. 295–313, *Exploding Wires*, vol. 3, Plenum Press, NY, 1964.

Nipher, Francis E. *Experimental Studies in Electricity and Magnetism*, 1914. (Printed by P. Blakiston's Son & Co., Philadelphia.)

Philosophical implications

Dreyfus, Hubert and Stuart Dreyfus. *Mind Over Machine: The Power of Human Intuition and Expertise in the Era of the Computer*. New York: The Free Press, 1986.

Elkin, Benjamin. *The Loudest Noise in the World*. New York: Viking Press, 1954.

Kuhn, Thomas. *The Structure of Scientific Revolutions*. Chicago: University of Chicago Press, 1996.

List of related contacts

The Bakken: A Library and Museum of Electricity in Life, 3537 Zenith Avenue South, Minneapolis, MN 55416. The Bakken has excellent research resources for scholars and many kinds of rare electrostatic apparatus on display.

The Burndy Library, c/o The Dibner Institute, 38 Memorial Dr., Cambridge, MA 02139. Houses and publishes books on the history of electricity. Excellent sources for scholars.

Electret Scientific Company, P.O. Box 4132, Star City, WV 26505. Publishes books on electrostatics applications, electret waxes, electrostatic motors and experiments.

Electrostatics Society of America, c/o Tim Erin, 723 Woodshire Way, Dayton, OH 45430.

The E.S.A., founded by Dr. A.D. Moore, presents awards for outstanding science fair projects, holds yearly conferences and publishes a newsletter. Open to anyone interested in the study of electrostatics.

The Charles Fort Institute, BCM Forteana, London WC1N 3XX, England.

Common Sense Science, Inc., P.O. Box 1013, Kennesaw, GA 30144-8013. Publishes a newsletter that reexamines the foundations of modern physics; proposes a vortex-ring model for the electron.

http://members.nbci.com/rford/

Institute for Frontier Science, 6114 LaSalle Ave., Box 605, Oakland, CA 94611.

Society for Amateur Scientists, 5600 Post Rd., Suite 114-341, E. Greenwich, RI 02818 (Director: Dr. Shawn Carlson).

Index

About the Author

R.A. Ford is an electrical experimenter and inventor specializing in turn-of-the-century electrostatic devices. An avid researcher, he developed his own electrostatic generator, which is detailed in the book. Ford also has served as a technical consultant to manufacturers of Wimshurst and Van de Graaff generators. For many years, he has devoted himself to introducing students of all ages to the science and history of electricity. For the past five years, he has worked to develop high voltage equipment for use in high school science fair projects as well as in physics labs and lectures.